L3 PE

S
540
.A2
.INV

INVESTMENT STRATEGIES FOR AGRICULTURE AND NATURAL RESOURCES

Investing in Knowledge for Development

Dedication

This book is dedicated to

Dr Ian Haines

formerly Deputy Chief Natural Resources Advisor,
Department for International Development, UK,

who was committed to the ideals of agricultural and natural
resources research to alleviate poverty, and died suddenly
at a meeting in pursuit of these ideals at FAO, Rome, on
16 December 1997. His friends who participated in this study
miss his wit and his wise counsel.

Investment Strategies for Agriculture and Natural Resources

Investing in Knowledge for Development

Edited by

G.J. Persley

CABI *Publishing*

in association with

The World Bank
Australian Centre for International Agricultural Research
UK Department for International Development
Deutsche Gesellschaft für Technische Zusammenarbeit

CABI *Publishing* – a division of CAB INTERNATIONAL

CABI *Publishing*
CAB INTERNATIONAL
Wallingford
Oxon OX10 8DE
UK

CABI *Publishing*
10 E. 40th Street
Suite 3203
New York, NY 10016
USA

Tel: +44 (0)1491 832111
Fax: +44 (0)1491 833508
Email: cabi@cabi.org

Tel: +1 212 481 7018
Fax: +1 212 686 7993
Email: cabi-nao@cabi.org

A catalogue record for this book is available from the
British Library, London, UK.

Library of Congress Cataloging-in-Publication Data
Investment strategies for agricultural and natural resources :
   investing in knowledge for development / edited by G.J. Persley.
      p.   cm.
   Includes bibliographical references (p.   ) and index.
   ISBN 0-85199-280-3 (alk. paper)
      1. Agriculture- -Research.   2. Natural resources- -Research.
   3. Agriculture- -Research- -Economic aspects.   4. Natural resources-
   -Research- -Economic aspects.   I. Persley, G.J.
   S540.A2I56   1998   98-8806
   338.1'3- -dc21   CIP

   ISBN 0 85199 280 3

Typeset in Melior and Optima by Advance Typesetting Ltd, Oxford
Printed and bound by Biddles Ltd, Guildford and King's Lynn

Contents

v

Contributors

Gary Alex, ESDAR, The World Bank, 1818 H Street NW, Washington, DC 20433, USA

J.M. Alston, Department of Agricultural and Resource Economics, University of California, Davis, CA 95616, USA

Alois Basler, Institute of Agricultural Market Research, Federal Agricultural Research Center, Braunschweig-Volkenrode, Germany

A. Beattie, UK Representative to the UN Food and Agriculture Agencies in Rome, Viale Aventino 36, 00153 Rome, Italy

N.P. Clarke, The Agriculture Program, The Texas A&M University System, Centeq Research Plaza, Suite 241, College Station, Texas 77843-2129, USA

M.C. Crawley, Foundation for Research, Science and Technology, 43 Tirohanga Road, Lower Hutt, New Zealand

J.J. Doyle, 45 St Germains, Bearsden, Glasgow G61 2RS, UK

E.F. Henzell, 182 Dewar Terrace, Corinda, Queensland 4075, Australia

R.W.M. Johnson, Consultant to New Zealand Ministry of Agriculture (MAF), 57 Allington Road, Wellington, District 5, New Zealand

A. Kissi, Institut National de la Recherche Agronomique, Rabat, Morocco

U. Lele, ESDAR, The World Bank, 1818 H Street NW, Washington, DC 20433, USA

M. McMahon, LAC Technical Department, The World Bank, 1818 H Street NW, Washington, DC 20433, USA

D. Mentz, Mentz International Trading and Investment Ltd, Unit 1, 25 Longridge Road, London SW5 9SB, UK

B. Nestel, Little Goldwell Oast, Goldwell Lane, Great Chart, Ashford, Kent TN23 3BY, UK

M. Noor, SPAAR Secretariat, The World Bank, 1818 H Street NW, Washington, DC 20433, USA

P. Pardey, Research Fellow, International Food Policy Research Institute, Washington, DC 20036, USA

G.J. Persley, Biotechnology Alliance Australia, University of Queensland, St Lucia, Brisbane, Queensland 4072, Australia

A.D. Portugal, EMBRAPA, Caixa Postal 040315, 70770-901 Brasilia DF, Brazil

A. Reguragui, Institut National de la Recherche Agronomique, Rabat, Morocco

F.J.B. Reifschneider, Secretariat for International Cooperation, EMBRAPA, Caixa Postal 040315, 70770-901 Brasilia DF, Brazil

C. Roturier, Association de Coordination Technique Agricole, 149 rue de Bercy, 75595 Paris cedex 12, France

G. Steinacker, G.S. GTZ-4233, P.O.B. 5180, D-65726 Eschborn, Germany

E.S. Wallis, Sugar Research and Development Corporation, PO Box 12050 Elizabeth Street, Brisbane, Queensland 4002, Australia

Foreword

In meeting the development challenges of the 21st century, against the background of a doubling of the world population and increasing urbanization, we shall need to develop better ways of managing natural resources while protecting the environmental assets of the planet *and* eliminating poverty.

At the heart of this agenda is the need for a continuing supply of new knowledge and better technology that will inform policy-making; increase production and productivity; provide a wide range of opportunities and options for development; conserve the environment; plus improve livelihoods and our ability to monitor the impact of development processes on natural resources and the environment.

Some of this work will be funded by the public sector, particularly where the outputs are intended to be public goods or common property; however, increasing amounts of work are, and will be, commissioned by the private sector and civil society organizations. Within this agenda there is scope for building productive and mutually beneficial partnerships.

There is a growing body of evidence of the returns that can be secured from investment in well-managed and well-targeted research. However, as demands on public sector finance increase, it is important that public resources are used efficiently and effectively. Many governments are exploring new structures and institutional relationships between the providers of research and knowledge services, and 'funders' or customers for technology and products, aimed at improving performance and promoting sustainable development.

From the analysis and discussion of approaches used it is clear that 'what works best' will vary with local and national circumstances. However, some common 'lessons learned' and 'good practice' have emerged:

- It is important to involve farmers and other key stakeholders or partners at all stages in the process – from the identification of priorities to the funding of work and dissemination of the results – and to focus on results.
- As the challenges get greater the need for new strategic alliances and interdisciplinary approaches at local, national and global levels becomes more necessary.
- In the development of new approaches, the sharing of experience should be continuous and the process evolutionary. We should make better use of new communication systems to encourage the sharing of experience and lessons learned.

This book starts this exchange. The dialogue and publication are evidence of growing international partnerships – in this case between the World Bank, the Australian Centre for International Agricultural Research, Deutsche Gesellschaft für Techische Zusammenarbeit (GTZ) and the Department for International Development (UK), and the many authors who contributed their time and experience.

I hope that it will stimulate interest and discussion, and encourage further exchanges.

Andrew Bennett, CMG
DFID, London, UK

Preface

A study was initiated in 1996 to examine the experiences and the lessons learned in the reform of public sector R&D in selected OECD and developing countries over the past decade. The purpose of the study was to identify any overarching lessons from the reforms in individual countries that are relevant to future support of national and international agricultural research by national governments and by international development agencies.

Status papers were commissioned from authors in several countries who have been involved in the substantial changes in policy, finance, execution and delivery of agricultural and natural resources research. Among industrial countries, these included Australia and New Zealand, France, Germany, the UK and the USA. The changing scenarios in support of agricultural and natural resources research were also reviewed regionally for Latin America and the Caribbean, sub-Saharan Africa and Asia. Case studies were presented for Brazil and Morocco. The economics of public sector investments in public sector R&D in agriculture were also reviewed, as were changes in the research agenda being wrought by increasing emphasis on natural resources management and the use of biotechnology.

The commissioned authors and other advisers were brought together at a workshop in London in October 1997 to identify the key lessons emerging from the study to date, and to discuss the next steps. These findings are being elaborated in a synthesis paper to be published by the World Bank. The commissioned papers, which give the details of the experience in individual countries and regions are published in this volume. The sponsors are also looking at ways to make a synopsis of the material available on the Internet and to facilitate the development of a web site linked to a knowledge management system which would allow the continual sharing of experience among policy-makers, research managers and researchers in the many countries and international agencies involved in the reform of agricultural and natural resources research.

Gabrielle J. Persley

Acknowledgements

The contribution of all authors to this volume is gratefully acknowledged. All who participated at the October 1997 workshop in London made significant inputs into the review of the papers and the discussion of the issues arising, and their comments and suggestions are reflected in the edited versions of the papers herewith and in the concluding chapter.

The support of ACIAR, DFID, GTZ and the World Bank who co-sponsored this study is gratefully acknowledged. I would like to thank especially the following persons who provided advice and encouragement both in the development of the concept and in the conduct of the study: Dr John Doyle, Mr Andrew Bennett of CMG, Dr Mike Scott, Ms Lucy Ambridge and especially the late Dr Ian Haines of DFID; Mr Michel Petit and Mr Gary Alex of the World Bank; Dr Ian Bevege of ACIAR; and Dr A. Springer-Heinze of GTZ. The participation of Dr Julian Alston and Dr Phil Pardey and his colleagues from IFPRI in the study also brought a wealth of economic data arising from their extensive work in the field of science policy. The work of ISNAR has also been particularly valuable, as reflected in their recent publication of *Financing Agricultural Research: a Sourcebook*, edited by Steven Tabor, Willem Janssen and Hilarion Bruneau.

Skilful administrative and organizational support for the London workshop was provided by Ms Lucy Ambridge of DFID and Ms Pamela George of the World Bank. Mr Reginald MacIntyre undertook the technical editing of the manuscripts and Ms Ida MacIntyre provided valuable backup editing and word-processing support. I am grateful to Mr Tim Hardwick, CABI Book Publisher, for his continued support, and to Rachel Robinson, CABI Editorial Coordinator. The contributions of all are gratefully acknowledged.

Acronyms and Abbreviations

ACIAR	Australian Centre for International Agricultural Research
ACTA	Association for Agricultural Technical Coordination (France)
ADB	Asian Development Bank
AIDA	Allianz für International Ausgerichtete Deutsche Agrarforschung (German Alliance for International Agricultural Research)
ANDA	National Agricultural Development Association (France)
ANZ	Australia/New Zealand
APAARI	Asia Pacific Association of Agricultural Research Institutes
ATSAF	Arbeitsgemeinschaft für Tropische und Subtropische Agrarforschung (Council for Tropical and Subtropical Agricultural Research)
AVRDC	Asian Vegetable Research and Development Center
BID	Inter-American Development Bank
CAP	Common Agricultural Policy (European Union)
CBD	Convention on Biodiversity
CETAs	Centres des Études Techniques Agricoles (France)
CGIAR	Consultative Group on International Agricultural Research
CIAT	Centro Internacional de Agricultura Tropical
CIMMYT	Centro Internacional de Mejoramiento de Maiz y Trigo
CIP	Centro Internacional de la Papa
CIRAD	Centre de Coopération Internationale en Recherche Agronomique pour le Développement

CIRST	Conseil Interministeriel de la Recherche Scientifique et Technique
CNRF	Centre National de Recherche Forestière (Morocco)
CRSP	Collaborative Research Support Program
CSIRO	Commonwealth Scientific and Industrial Research Organization (Australia)
CTIFL	Intervocational Technical Centre for Fruits and Vegetables (France)
DFG	Deutsche Forschungsgemeinschaft (German Union for Scientific Research)
DFID	Department for International Development (UK)
DSIR	Department of Scientific and Industrial Research (NZ)
EMBRAPA	Empresa Brasiliera de Pesquisas Agropecuarias (Brazil)
ENFI	École Nationale Forestière d'Ingenieurs de Sale (Morocco)
ESDAR	Agricultural Research and Extension Group of The World Bank
EU	European Union
FAO	Food and Agriculture Organization of the United Nations
FNDA	National Agricultural Development Fund (France)
FNSEA	National Federation of Farmers Unions (France)
GARS	Global Agricultural Research System
GATT	General Agreement on Tariffs and Trade
GDP	gross domestic product
GIS	Geographic Information System
GNP	gross national product
GTZ	Deutsche Gesellschaft für Technische Zusammenarbeit (German Agency for Technical Cooperation)
IARCs	International Agricultural Research Centres
IBSRAM	International Board for Soils Research and Management
ICIMOD	International Centre for Integrated Mountain Development
ICRAF	International Centre for Research in Agroforestry
ICRISAT	International Crops Research Institute for the Semi-arid Tropics
ICTAs	Instituts et Centres Techniques Agricole (France)
IDRC	International Development Research Centre
IFPRI	International Food Policy Research Institute
IICA	Inter-American Institute for Co-operation in Agriculture
IIMI	International Irrigation Management Institute
IITA	International Institute of Tropical Agriculture
ILRAD	International Laboratory for Research on Animal Diseases
ILRI	International Livestock Research Institute

INIA	National Institutes for Agricultural Research (LAC)
INRA	National Agricultural Research Institute
IPGRI	International Plant Genetic Resources Institute
IPM	integrated pest management
IPR	intellectual property rights
IRRI	International Rice Research Institute
ISNAR	International Service for National Agricultural Research
ITCF	Technical Institute for Grains and Fodder (France)
KARI	Kenyan Agricultural Research Institute
LAC	Latin America and the Caribbean
LGU	Land Grant University (USA)
NARS	National Agricultural Research System
NGO	non-governmental organization
NRMR	natural resources management research
OECD	Organization for Economic Cooperation and Development
ORSTOM	Office de Recherche Scientifique et Technique Outremer
OTA	US Office of Technology Assessment
R&D	research and development
RIRF	Rural Industry Research Funds (Australia)
SACCAR	Southern African Center for Cooperation in Agricultural and Natural Resources Research and Training
S&T	science and technology
SAES	State Agricultural Experiment Station (USA)
SNPA	Sistema Nacional de Pesquisa Agropecuaria (Brazil)
SPAAR	Special Program for African Agricultural Research (World Bank)
SSA	sub-Saharan Africa
TAC	Technical Advisory Committee (of the CGIAR)
TRIPS	Trade-related Aspects of Intellectual Property Rights
UNDP	United Nations Development Programme
UNEP	United Nations Environment Programme
USAID	United States Agency for International Development
USDA	US Department of Agriculture
WARDA	West African Rice Development Association
WB	World Bank
WTO	World Trade Organization

Research Investment Strategies in Selected European Countries: Lessons Learned

I

The United Kingdom*

Anthony Beattie

*UK Representative to the UN Food and Agriculture
Agencies in Rome, Viale Aventino 36, 00153 Rome, Italy*

Introduction

This chapter reviews UK experience in public sector science policy
and management in the period 1971–97 (and accordingly does not
reflect developments since mid-1997). It is directed at practitioners
who want to discover quickly and efficiently whether the British experience has anything to offer them in (re)designing and managing
research systems, and where to turn for further information. This
approach affects both the content, which is deliberately selective, and
the presentation, which leans towards the telegraphic rather than the
discursive.

The chapter does not deal specifically with R&D in the agriculture
and natural resources sector. The models and lessons that are potentially
relevant elsewhere can in principle be found in any sector, and a limita-
tion to agriculture and natural resources would accordingly be artificial.

The UK's scientific and technological capability is made up of three
main inter-independent elements:

1. *The science, engineering and technology base*: the knowledge,
research and problem-solving skills, and the equipment and facilities
in universities and Research Council institutes. This is funded mainly
by the Office of Science and Technology (OST), which is part of the
Department of Trade and Industry, and the Education departments.

*The views expressed in this chapter are not necessarily shared by the
UK Government.

©CAB INTERNATIONAL 1998. *Investment Strategies for Agriculture and
Natural Resources* (ed. G.J. Persley)

2. *The technological and innovation capabilities of UK-owned and UK-based firms*: their workforce skills, knowledge base, stock of equipment and facilities and ability to access and absorb technology.
3. *The technological capabilities of research and technology organizations, of government laboratories and research facilities, and of other non-commercial bodies.*

The UK invests about £14 billion (US$22.6 billion) a year on R&D or 2.2% of GDP. Some £9 billion (US$14.5 billion) of the total is funded by the private sector and £5 billion (US$8.1 billion) by the public sector. The three main components of public sector science and technology (S&T) are:

1. Funding by the Ministry of Defence (£2.2 billion, US$3.5 billion), largely in support of equipment procurement; £1.6 billion (US$2.5 billion) is spent on development, the remaining £0.6 billion (US$1 billion) on research.
2. Funding by civil departments (£1.1 billion, US$1.8 billion) in support of their policy and regulatory functions.
3. Spending on basic and strategic research in universities and Research Council Institutes: the science and engineering base. Total spending on the science base is some £2.3 billion (US$3.7 billion). This comprises £1 billion (US$1.6 billion) from the Education departments to support the research infrastructure in universities (mainly academic salaries and equipment) and £1.3 billion (US$2.1 billion) spent by the Research Councils on projects undertaken for them by universities, on research in the Councils' own institutes, on national research facilities and on subscriptions to major international facilities such as CERN.

Since 1987 there have been significant changes in the pattern of public sector spending on S&T:

- Defence spending on R&D has fallen by 36% in real terms and is planned to fall further.
- Spending by civil departments has also fallen by 46% in real terms. Most departments have been affected, but the largest reductions have resulted from the ending of the fast breeder programme.
- In contrast, spending on the science base has grown by 10% in real terms, but on present plans (at time of writing, i.e. mid-1997) is now declining.

Although overall government S&T expenditure is on a declining trend, private sector expenditure is rising. It presently accounts for 67% of the UK's total investment in R&D, compared with 57% 10 years ago. But private sector spending still lags behind many of the UK's competitors, both as a proportion of GDP and on other indicators such as the ratio of R&D to sales or turnover. The UK's total (public and private) investment in R&D currently ranks fifth in the world (2.2% of GDP), behind France (2.4%), Germany (2.4%), the USA (2.5%) and Japan (2.7%). Total UK spending on defence R&D (public and private sectors taken together) has declined sharply and is continuing to decline, but still accounts for some 15% of the total. In this respect the UK is similar to France (17%) and the USA (20%). In Japan and Germany, by contrast, civil R&D predominates. Finally, it should be noted that agriculture, horticulture, forestry and fishing contribute around 1.5% to GDP.

Policy Landmarks

The period since 1971 has been marked by continuous change in science policy and management, accompanied by a vigorous public debate. Underlying the changes have been: (i) a strong wish on the part of successive governments to harness the UK's scientific capacity more effectively for the purposes of wealth creation; (ii) a drive for greater effectiveness and efficiency in the use of public funds; and (iii) a more restrictive view of the proper role of the public sector. The principal landmarks are outlined below.

The customer–contractor principle

A government green paper (consultation document) was published in 1971 (Anon., 1971), which came to represent a watershed in national science policy. It included reports by Lord Rothschild, in his capacity as head of the Central Policy Review Staff, and by a Working Group of the Council for Scientific Policy chaired by Sir Frederick Dainton. In his report, indisputably the more influential of the two, Rothschild enunciated his now famous customer–contractor principle as the most efficient way of procuring applied R&D: 'the customer says what he wants; the contractor does it (if he can); and the customer pays.' (He went on to say that 'basic, fundamental or pure research ... has no analogous customer–contractor basis.')

In July 1972 the British Government announced that Rothschild's recommendations were to be put into effect, as follows:

- The customer–contractor principle would be applied to all government applied R&D, whether carried on inside or outside government departments.
- The scientific capabilities of departments would be strengthened, to enable them to implement the approach and to provide scientific advice on departmental needs and policy formulation.

- A phased (3-year) transfer of part of the funds provided by the government for the Agricultural Research Council, the Medical Research Council and the Natural Environment Research Council to departments, to enable them to commission applied work. No restrictions were placed on the use of the money transferred to customer departments, but the expectation was that it would be used to procure work from the Research Councils.
- Contractors could in principle levy a 10% surcharge on work commissioned with them in order to finance innovative or longer-term research of their own choosing.

These changes, implemented at varying speeds across the public sector, had a profound effect on relationships within the scientific community and between the scientific community and its new providers of funds. Unsurprisingly, scientists were seriously discontented by the shift of power to 'customers', whom they regarded as generally ill-equipped, scientifically or managerially, to control the resource allocation process. The debate about the validity of the customer–contractor principle continues to this day, though in recent years it has become more muted.

During the 1980s this approach to the financing of R&D became swept up in the wider and deeper process of public sector reform, in which essentially similar principles were applied (among other places) to the health and education sectors. What has been called the 'dethronement of the professionals' in these sectors produced feelings of discontent similar to those expressed earlier by scientists. Over the years, however, the principal features of the 'new public management' – the creation of markets or quasi-markets, the separation of policy functions from service delivery and the provision of incentives to organizations and individuals – have become the conventional wisdom; they now face no serious challenge from the scientific community.

The retreat from near-market research

The British Government announced in 1988 that public expenditure on R&D would henceforth be more strictly governed by the market failure principle, i.e. that the government should only intervene where the market, left to its own devices, would fail to optimize benefits for the economy as a whole. It followed that public funds should be confined to activities in the basic and strategic parts of the R&D spectrum: the development of a marketable product or process – 'near-market research' – was the proper responsibility of industry. This policy resulted in the phasing out of most government support for near-market and single company R&D in the civil sector.

The internal market for government R&D

In 1989 the Chief Scientific Adviser to the government (Sir John Fairclough) restated the Rothschild customer–contractor principle as the basis for procurement of scientific services by government departments. He promulgated certain rules intended to produce an 'internal market' across the public sector under which public sector research suppliers would compete on fair terms for such services. The principal rule in what came to be known as the Fairclough Guidelines was that departments should employ competitive tendering procedures for placing S&T work unless there were significant cost or wider value-for-money considerations which justified an alternative approach. Departments were henceforth required to report progress annually towards the liberalization of the market.

Institutional change

In 1988 the British Government published a stock-taking review on civil service reform since 1979 (Anon., 1988), which was to have a significant effect on the operation and ultimately the ownership of government research establishments. (More widely it had an impact on the civil service in general, more profound than anything since the Northcote–Trevelyan reforms of the 19th century.) In response to a recommendation that discrete government functions should be reconstituted at arm's length from their parent departments, all 12 government research establishments were progressively established as 'agencies'.

Agencies have a large measure of autonomy under a chief executive reporting direct to ministers, and are managed under a regime that requires them to recover their expenses from the sale of services priced at full economic cost. The creation of agencies represented a step change towards the commercialization of the relationship between departments and their in-house research establishments, and put some teeth into this for the first time (though not enough according to the Stewart–Levene Report, 1992). Although not intended at the time, it is now clear that the reconstitution of government research establishments as agencies was an important precursor to the privatization of some of them.

In the second half of the 1980s there were parallel changes in the Research Councils as new managerial approaches were introduced and institutional structures rationalized and overhauled. The result was extensive closure and consolidation of facilities, with accompanying reductions in staff; reorientation of scientific programmes; the separation of funding into directed and responsive modes; and a shift in the balance of funding away from the Councils' own institutes towards the higher education sector. When it came to review the scene in 1992, the Stewart–Levene Report was rather more impressed with the vigour and commitment shown by the Research Councils in their pursuit of value for money than it was with government departments and their research establishments.

Higher education

The funding of research in higher education institutions (HEIs) has been the subject of a number of important changes over the period:

- From 1990 to 1991 the University Funding Council separated the funds it provided for research from those provided for teaching. Since then, within a single block grant for teaching and research, the research component has been allocated with regard to the quality of research in individual HEI departments, assessed retrospectively under quinquennial peer reviews (research assessment exercises or RAEs). The research–R–grant is not earmarked for particular activities. Its purpose is to finance research of the institutions' own choosing and to support the infrastructure for projects undertaken for Research Councils under the dual support system described below.
- The 1991 White Paper on Higher Education (Anon., 1991) announced the abolition of the binary line between universities and polytechnics (as a result of which all polytechnics have now acquired the status of universities) and the reconstitution, effective from 1993–1994, of the respective funding bodies as funding councils for England, Wales and Scotland.
- The 1990 Autumn Statement on public expenditure amended the long-established dual support system for channelling research funds to HEIs, by switching a proportion of funds to Research Councils, which then became responsible for all the costs of research carried out for them in HEIs, except for academic salaries and premises.

The 1993 White Paper on Science and Technology (Anon., 1993b) left the dual support system untouched (despite an evident temptation on the part of the Stewart–Levene Report (1992) to replace it with an arrangement more in line with market principles). The main effects of the changes outlined above have been:

- A shift in the balance of funding from the Funding Council to the Research Council stream of the dual support system, with the growth in Research Council funds more than offsetting the decline in money from the Funding Councils.

- Increased selectivity in the allocation of research funds from both sources.
- A shift of emphasis from basic, curiosity-driven research towards goal-oriented funding.
- A shortfall in investment in research equipment and laboratories.
- A shift in the employment status of university researchers, with the number of short-term contract research staff more than doubling between the late 1970s and the middle 1990s.

The 1997 *Report of the National Committee of Enquiry into Higher Education* (the Dearing Report, HMSO) has propounded the following six principles for research funding in higher education:

1. Excellence should be supported.
2. Adequate funding should be provided for infrastructure to support high quality research and training.
3. Whatever research is chosen for support must be fully funded.
4. Funding policies for research should so far as possible promote, not devalue, teaching.
5. Different types of research should be supported by different streams of funding, including support for applied and regional work.
6. Funding streams and mechanisms should be clear and transparent.

The report went on to recommend:

- That projects and programmes funded by the Research Councils should meet their full indirect costs, including premises and central computing.
- That the next RAE for universities should be amended so as to encourage institutions to make strategic decisions whether to enter departments for competitive funding or to seek a lower level of non-competitive funding to support research and scholarship which underpins teaching.
- That government should establish an Industrial Partnership Fund to attract matching funds from industry and to contribute to regional and economic development.
- That government should facilitate the establishment of a revolving loan fund, financed jointly by public and private sector sponsors, to support infrastructure in a small number of top quality research departments that can demonstrate a real need.
- The setting up of a high level independent body to advise government on the direction of national policies for public funding of research in higher education institutions, on the distribution and level of such funding, and on the performance of the public bodies responsible for distributing it.

The 1993 White Paper on S&T

In May 1993 the government published a white paper (Anon., 1993b) spelling out a strategy for science, engineering and technology. This followed the first general review of policy and organization since the reports by Rothschild and Dainton (Anon., 1971). The White Paper was significantly influenced by the findings of a public consultation and drew extensively on a specially commissioned report, *Review of Allocation, Management and Use of Government Expenditure on Science and Technology* (Stewart–Levene Report, 1992). The public consultation pointed to the following:

- A widely perceived contrast between UK excellence in science and technology and the country's relative weakness in exploiting this to economic advantage.
- The absence of a clear statement of government objectives, which led to mixed and sometimes contradictory statements to the scientific and engineering communities.
- Within the context of limited resources, the need to manage government investment in science and technology to better effect.
- A need for more effective mechanisms for implementing policy, including policies related to international collaboration.
- Problems over the management of careers in science and engineering.

The aim of the Stewart–Levene Report (1992) was to make recommendations that would: (i) improve the effective allocation and efficient use of public funds committed to S&T; and (ii) produce better coordination of S&T policy across government departments. It recommended the following:

- Overarching statement of government objectives and priorities for S&T in the White Paper.
- Better machinery for reviewing and coordinating expenditure on S&T across government, including an enhanced role for the government's Chief Scientific Adviser in the annual Public Expenditure Survey, and concordats between Research Councils and departments involved in S&T.
- Annual departmental reviews of S&T policies and spending, published statements of strategy and an overview role for each department's chief scientist.
- Accelerated progress towards an open market for government S&T spending, with all research providers, whether government research establishments, Research Council institutes, universities or private sector bodies, eligible to compete for business.

- The separation of ownership and customer roles (on the grounds that these could not be successfully combined in a single department), either by privatizing government research establishments or by reorganizing them into one or more civil research agencies independent of their main customers.
- Removal of the restrictions that prevented government research establishments from diversifying their markets.
- Active private sector involvement in decisions on S&T expenditure directed at wealth creation, and in the work itself.
- New missions and structures for the Research Councils, reflecting government priorities for S&T and the importance of user involvement.
- The introduction of accounting systems which would enable universities to track the use of the research–R–grant from the Funding Councils and to provide the basis of recovering full economic costs for contract research.
- No change to the dual support system for funding research in universities.

The White Paper announced the government's commitment to harness S&T more effectively in the cause of national wealth creation and improving the quality of life. To carry this into effect the government would:

- open up the flow of information and debate on its S&T strategy, notably through the publication of an annual Forward Look with a 10-year horizon, which reviews plans for S&T investment across the public sector against White Paper objectives;
- launch a Technology Foresight programme, jointly with industry and the science and engineering communities;
- reorganize the machinery for consultation on science policy;
- redraw the boundaries of the Research Councils and give them new missions reflecting a commitment to wealth creation and the quality of life;
- review and improve schemes for technology transfer;
- maintain and strengthen the customer–contractor machinery used by departments to procure applied R&D;
- privatize as many government research establishments as possible and review the management regimes of those that had to remain in public ownership;
- maintain the dual support system for funding research in universities; and
- develop new arrangements for the postgraduate training of scientists and engineers.

The efficiency scrutiny of public sector research establishments

As noted above, the government signalled (Anon., 1993b) its intention to look again at the scope for privatizing research establishments owned by government departments and to review the management of those that could not be privatized. This reflected three things: (i) a more robust general policy on privatization of central government functions; (ii) a concern that the progressive development of open contracting for scientific services by departments was a wastefully slow way of determining whether there were – as some people suspected – too many such establishments; and (iii) a view that privatization was ultimately the most satisfactory way of achieving the efficiency gains that would flow from the separation of owner and customer roles recommended by the Stewart–Levene Report (1992).

The task was tackled by means of an Efficiency Scrutiny which reported in 1994 (Cabinet Office, 1994). After a period of public consultation the government announced in September 1995 (HMSO, 1995) that it did not accept the main findings and proposed to revisit the issues of ownership and management in a coordinated programme of 'prior options reviews'. The programme coverage was extended to include not just departmentally owned research establishments but also most Research Council institutes and other non-departmental public bodies engaged in research (e.g. the National Radiological Protection Board). The government's decisions on the reviews were the subject of a series of announcements in the period May 1996 to February 1997. Taken together with earlier decisions, the upshot was the privatization (in one case the contractorization) of seven major establishments out of a total of around 40 surveyed and the launch of a coordinated programme to improve the effectiveness and efficiency of establishments that remained in the public sector. (It is worth noting that four of the seven privatized establishments were management buyouts supported by private sector capital and that the transfer of the fifth was management-inspired.)

Treasury study of basic science

In July 1996 the UK Treasury published an important review of academic literature commissioned from the Science Policy Research Unit at the University of Sussex (SPRU, 1996). The review pointed to six main benefits from basic research:

1. Basic research as a source of *new useful information*.
2. The creation by basic researchers of *new instrumentation and technologies*.

3. *Skills* developed by those engaged in basic research – especially graduate students – which yield economic benefits when individuals move from basic research carrying codified and tacit knowledge.
4. Participation in basic research to gain *access to networks of experts and information.*
5. Skills in *solving complex technological problems,* particularly in industry, by those trained in basic research.
6. The creation of *spin-off companies.*

The study noted that a comparatively small proportion of the benefits flows in the form of new useful knowledge directly incorporated in new products or processes, and that there is great heterogeneity in the relationship between basic research and innovation as between scientific field, technology and industry. This complexity adds to the problems of trying to devise an optimum structure for government support of basic research. The emphasis on the economic benefits flowing from skill development, including the 'entry ticket' to networks, echoes studies elsewhere on development of S&T capabilities in Japan and other Asian economies.

Tools for Planning and Management

In the period covered by this review, British policy-makers and managers have developed an extensive shelf of tools for handling issues ranging from the determination of national science policy to the privatization of research establishments and the procurement of scientific services from the open market. The following paragraphs outline the main tools and models which may have some application in other systems.

Strategic choice: technology foresight

The White Paper's (Anon., 1993b) central initiative for bringing business people and scientists and engineers together to discuss market and technology trends was the national Technology Foresight programme, launched at the end of 1993. The programme aimed:

- to generate a 'bottom up' consensus about national and global market opportunities over the next 10–20 years;
- to identify the underlying science, engineering and technology investments that will help business to exploit those opportunities;
- to foster new and better networks between business, the science and engineering communities and government: this is seen as a key cultural change.

The work has been carried out by 15 sector panels. Between them the panels have made over 300 recommendations, relating not just to science, engineering and technology priorities, but also to educational, financial, regulatory and organizational issues. Many of the priority areas do not map directly on to science-based disciplines. They depend on contributions from a (sometimes very large) number of disciplines and the routes by which fundamental science contributes to them are complex and may often occur over a long period of time. A striking feature of the findings is the dependence of future competitiveness on a wide range of factors. These include proactive, technologically aware management, a suitably skilled workforce, a well-informed grasp of customer preferences and attitudes, effective transfer and exploitation of new technologies and social, economic and regulatory policy frameworks that are conducive to innovation.

A Technology Foresight Steering Group has produced an overarching report which identifies some 30 generic or cross-sectoral areas of science, engineering and technology (including 11 that warrant early attention by funding bodies on grounds of outstanding market opportunity, pervasiveness and UK capability) and some 20 priorities relating to infrastructural issues. The programme has provided, for the first time in the UK, an authoritative and forward-looking analysis of market needs across the whole economy. Its findings are being widely used to inform judgements about priorities, both within government departments and within the science base. It is too early to judge whether the pay-off will justify the effort that has been invested, but the programme has been widely welcomed as marking a step change in the approach to strategic choice in S&T, and in the creation of dialogue between the key stakeholders in the public and private sectors.

Strategic choice: the annual Forward Look

A second central aim of the 1993 White Paper (Anon., 1993b) was that the strategy of focusing the country's capabilities in science and engineering on the creation of wealth and improvements in the quality of life should be monitored and evaluated through an annual Forward Look process. The Forward Look sets strategic objectives over a 5–10 year perspective, taking into account:

- gaps in the education, training and research effort, set against the changing economic, social and scientific environments and considered alongside the activity of the private sector, including the research charities;
- how UK S&T compares with that of the country's principal competitors;

- the balance between civil and defence research, and between civil research commissioned by departments and that undertaken by the science and engineering base;
- the balance between domestic and international research and the scope for cooperation with other countries;
- opportunities for achieving greater synergy across programmes;
- the scope for more concerted action and collaboration, both within the public sector (between Research Councils, universities and government departments) but also between the public and private sectors.

The Forward Look is published as a single document, written for a broad readership but in particular for the industrial and research communities. It is supported by more detailed strategy and programme statements published by government departments, by corporate plans published by the Research Councils and by reports published by other public bodies such as the Higher Education Funding Councils.

How to be a good customer for research

The Stewart–Levene Report (1992) took as its starting point the principle that value for money was most likely to be secured if decisions about the management and use of S&T were taken by well-informed customers acting on the basis of clear objectives and with freedom of choice. The authors distinguished three kinds of public sector customer:

1. *Real or direct customers,* where government is the consumer, e.g. research to inform policy development or as part of the process of procurement.
2. *Proxy or indirect customers,* where government represents a wider group. The research might involve the development of a generic technology that no single company would invest in because of market failure, or public goods such as measurement standards or clean air. Much of the mission-oriented work of the Research Councils is of this kind.
3. *Sponsors,* where public funds are used to support curiosity-driven research for which objectives are not specified by the funder. The main examples are research financed from the R component of the Funding Councils' block grants to universities and the responsive mode activities of the Research Councils.

It is widely (though not universally) accepted that research in the first and second categories is best managed according to the customer–contractor principle enunciated by Rothschild (Anon., 1971). The

Stewart–Levene Report (1992) argued that best practice management of such research by a well-informed customer involves six things:

1. *Objective setting and output definition*: the customer should specify what he/she wants to buy in terms of outputs and testable objectives.
2. *Flexibility*: the customer should not be constrained in his/her choice of research contractor (unless by choice, in which case he/she must be able to justify the choice).
3. *Selectivity*: the customer should allocate funding on the assessed quality of research contractors or teams.
4. *Cost control*: the customer should be able to get value for money.
5. *Monitoring*: regular exchange of information between customer and contractor, with the customer free to redirect the work if necessary.
6. *Evaluation*: critical evaluation of outputs against objectives established at the outset, with the results lodged in the corporate memory.

In its evaluation of the performance of the three customer groups identified above, the Stewart–Levene Report (1992) included the following points:

- Criticism of government departments for a slow and patchy application of the customer–contractor principle and an elastic interpretation of the Fairclough Guidelines on the Internal Market. In part this was attributed to cosy relationships between government departments and their in-house research establishments. The Report concluded that the responsibilities of ownership would generally blunt the rigour with which customers should manage their contracts and therefore recommended separation of the two roles.
- The report was generally impressed with the Research Councils' approach to directed and responsive mode funding. This involved priority setting in annual corporate plans, appraisal by means of peer review and a willingness to close and restructure programmes in the interests of relevance and cost-effectiveness. What was needed now was a clearer specification via their missions of what government wanted the Research Councils to do, greater involvement in decisions and research by the user community and a willingness to deploy funds outside the Science Base.
- Criticism of the accountability arrangements in the Funding Councils' system for distributing research grants to universities, on the one hand because the grant as such was not accounted for independently of the teaching component, and on the other because there was no information on the main purposes for which it was used. (The Report considered but rejected the idea of transferring some or all of the R grant to the Research Councils.)

Government departments now routinely publish their research strategies and identify and manage programmes and projects according to the ROAME (rationale, objectives, appraisal, monitoring and evaluation) methodology, or something resembling it. (The Department for International Development (DFID) uses a logframe approach.) Some departments have produced guidance manuals for their customer divisions which are in the public domain (e.g. the *Research Management Process Note 1996* published by the Office of the Chief Scientist in the Department of the Environment). The Research Councils operate planning and management systems on broadly similar principles, but with a particular focus on the peer review and visiting group arrangements which lie at the core of their appraisal and evaluation approaches. These are in some cases quite elaborate. Similarly there are extensive guidance papers from the Higher Education Funding Councils on the administration of the quinquennial Research Assessment Exercise (RAE).

How to get the supplier base right

The programme of public sector reform that was pursued by the government between 1979 and 1997 had a considerable impact on the ownership, scope and management systems of public sector scientific establishments. Many of the resulting system changes – which ranged from corporate planning to financial regimes, performance-related pay and benchmarking – were applied across the public sector, usually to government departments in the first instance and non-departmental public bodies (such as Research Councils) in the second. The issues that have turned out to be specific to (or to relate with particular force to) public sector research establishments largely revolve around two interrelated questions: (i) How to reconcile continued public sector status with being a supplier of scientific and technical services under customer–contractor arrangements; and (ii) The case for privatizing a particular establishment. The issues arising under the first have in many cases informed and clarified those arising under the second.

As a consequence partly of the customer–contractor arrangements which were put in place following the Rothschild report (Anon., 1971), but more importantly because of the quasi-commercial basis on which they were reconstituted as agencies after the *Next Steps* report (Anon., 1988), the dozen or so government research establishments owned by departments found themselves during the 1990s developing important new perceptions of the organizations as businesses. These perceptions revolved around:

- their heavy investment in assets, both physical and human, which imposed high costs and limited flexibility;

- a cost base dominated by pay;
- reliance on one or a few customers, which made the business highly vulnerable, coupled with a government-wide policy that restricted any initiatives to develop new markets;
- the value of business being bound up with intellectual capital, which put a premium on attracting and retaining first-rate people;
- an asymmetric risk profile, with a heavy downside risk and not much upside (easy to lose support from existing customers but hard to replace this elsewhere);
- a lack of intellectual freedom, with the research agenda essentially determined by customers in the parent department.

At the same time there were concerns at the centre of government that the Internal Market for scientific services was not working as intended, and that the continuing close relationships between departments and their research establishments were stifling the exploitation of efficiency gains, including those stemming from rationalization of facilities. The Stewart–Levene Report (1992) identified the following impediments to a more open market:

- The value placed on continuity and coordination of work.
- The cash cost of change (because departments were owners as well as customers they could not escape the financial consequences of, for example, redundancies if they decided as customers to place work elsewhere).
- The cost of running competitions (i.e. transaction costs).
- 'Level playing field' issues: departments were unwilling to open up markets unilaterally, fearing that their research establishments might be denied opportunities to compete for business elsewhere; coupled with concerns about whether prospective bidders for work (universities in particular) would price their services fairly.
- Perceived risks to the excellence of the establishments, in particular whether owner/customer departments would continue to invest in the development of new capabilities, and whether the most stimulating work would be placed elsewhere.
- Reliance on owned suppliers for independent advice on policy advice (or procurement advice in the case of defence).
- A need for high levels of responsiveness, e.g. in relation to food safety emergencies.
- Preservation of unique skills or facilities.

The Stewart–Levene Report (1992) concluded: (i) that there was substantial unexploited scope for efficiency gains as a result of open contracting; and (ii) that the best way to circumvent the problems inherent in the combined roles of ownership and customer was to separate the ownership of S&T facilities from the procurement of S&T

services. The separation was to be achieved ideally by privatization and, where that was not possible, by collecting establishments together into one or more civil research agencies. The benefits of separation were expressed as the following:

- An improved ability to *rationalize capacity* with a view to: (i) achieving more output for a given level of expenditure; (ii) bringing together currently fragmented capabilities; and (iii) creating greater flexibility to reshape public expenditure on S&T.
- Better *value for money for customers*, who would have the freedom to award work solely on supplier merit.
- A *wider range of opportunities for research providers* to forage more widely for work, including with industry, and to improve their competitiveness by shedding non-core activities.

These ideas were reflected in the 1993 White Paper (Anon., 1993b) and subsequently tested in the 1994 Efficiency Scrutiny and its associated programme of 'prior options' reviews of public sector research establishments. The methodology developed for the programme of prior options reviews represents the most advanced thinking on how to test the appropriateness and feasibility of privatizing public sector research establishments. It requires rigorous answers to the following questions:

1. Are the outputs (of establishment X) needed?
2. Must the public sector pay for some or all of them?
3. Does the public sector have to own some or all of the delivery mechanism?
4. If not, how feasible is it to extend private sector involvement in delivery by privatization; or failing that by contracting out?

To the extent that it is inappropriate or unfeasible to extend private sector involvement:

5. What is the scope for securing efficiency gains by rationalization?
6. How should the function(s) be managed in the public sector so as to maximize effectiveness and efficiency?

Arguably (1) and (2) should be conflated, since they have to do with market failure: they might therefore be better expressed as 'What is the case for public sector involvement in the supply of these outputs?' And (5) is both logically a subset of (6) and an issue that needs to be considered under (4). The phrase 'some or all' in (2) and (3) is particularly important. Since many public sector research establishments encompass more than one function, the scope for unbundling the functions and dealing with them in different ways needs to be considered. This means distinguishing functions and outputs clearly and looking carefully at the interdependencies and scope for unbundling.

Outcomes, Dilemmas and Issues for Developing Countries

Outcomes

The principal outcomes of the changes described above are as follows:

- Fewer resources are being invested in publicly funded S&T.
- It is accepted that the UK cannot be pre-eminent in all research fields and that public funds for research must therefore be invested on a selective basis.
- In an effort to promote relevance and value for money there has been a decisive and irreversible shift towards management by contract. Outside the higher education sector researchers have moved from a world of core grants to contracts for performance, awarded to an increasing extent competitively and based in many (but not all) cases on full economic costs.
- As a result of funding reductions, rationalization and privatization there are now substantially fewer publicly owned research institutes, and staff numbers have fallen significantly.
- A powerful battery of planning and management techniques has been developed, ranging from Technology Foresight through programme and project identification and management to institutional management issues.

Informed opinion would generally accept that the UK has substantially improved its understanding of the role and function of science in a mature economy, and that the investment of public funds in the sector is better directed and more cost-effective. But these benefits have been bought at a high price, including widespread discontent in the scientific community, many members of which remain critical of the way the sector has been treated over the period under review.

Dilemmas

Debate continues over many issues, notably:

- Whether the government is devoting sufficient money to science.
- Whether industry is investing appropriate resources in R&D.
- The effectiveness of technology transfer mechanisms between the public and private sectors and of machinery for commercializing university research and associated scientific skills.
- The scope for further privatization of publicly owned research institutes.
- The pace at which public procurement of scientific services (by government departments and Research Councils) should be further opened up to competition.

- How to produce a level playing field on which potential public and private sector bidders for research contracts and grants can fairly compete.
- How to ensure that the transaction costs which arise in the markets and quasi-markets that have been created do not become disproportionately high.
- Whether the research funding mechanism for universities needs to be rethought in order to protect the flow of resources to first-rank, research-led universities.
- How to find resources to modernize the research infrastructure of universities, now a subject of serious concern.
- How to ensure that the quality of science graduates from universities is appropriate to an increasingly knowledge-based economy.

Issues for developing countries

Against this background, the main lessons for policy-makers in developing countries seem to be as follows:

1. Develop early a clear vision and set of objectives for the sector, if possible in the context of a national policy for S&T. Where do you want to be and how will you recognize success?

2. Be realistic about the timescale for change. The first priorities for policy-makers are to get the objectives and architecture right, and to identify and manage the stakeholders.

3. Ensure that the stakeholders in publicly funded science have a clear understanding of the rationale for devoting public money to science. Without this there is scope for endless confusion and acrimony.

4. Consider the merits of a Technology Foresight (or some such) exercise as a way of establishing priorities and – particularly – engaging stakeholders.

5. Be selective about domestic investment in science as opposed to buying in from elsewhere (or free riding on the investments of others). But keep in mind the importance of suitably skilled scientists as an entry ticket to the international network and the opportunities it offers to exploit knowledge and skills from elsewhere.

6. Beware of 'producer capture', i.e. allowing existing institutional capacity and skills to drive the resource allocation process. Consider establishing arm's length relationships to protect the integrity of the decision-making process. These could range from a customer–contractor relationship between a research funding body and an in-house supplier, operated on the basis of single tenders, to a fully competitive model in which the research funder puts all its work out to open competition.

7. Be careful to ensure that the creation of a customer–contractor model for procuring scientific services does not squeeze out investment in long-term research by putting all the funds in the hands of customer departments whose horizons are necessarily short term.

8. Consider whether the publicly owned network of institutions needs rationalization and/or divestment. Look critically at arguments for continued public ownership, keeping in mind that these are different from the arguments for public *funding*. If there is a case for change there are two possible approaches: *laissez-faire*, allowing the market to decide which institutions should survive; and *dirigiste*, involving administrative action to close, rationalize or divest. The UK has tried both. The latter is infinitely to be preferred.

9. Do not reinvent the wheel. The effort that has been devoted to science policy and management issues by the UK and others over the past 20 years means that there is a raft of tools that can be imported and adapted as necessary.

10. In the interests of efficiency, ensure that scientific bodies that remain under public sector ownership are managed according to best practice principles (on which, again, there is a vast amount of transferable knowledge and experience).

11. Aim to push decisions downwards and outwards. Take decisions at lowest possible levels, on the basis of clear accountability for results. Be prepared to take risks in the interests of quick and efficient decision making.

12. Remember that the chief asset of an S&T system is people. Striking the right balance between control (in the interests of efficiency) and freedom (in the interests of creativity and innovation) is a difficult but crucial managerial challenge.

13. Keep a close watch on the sources of supply of scientific personnel for the sector. Problems with numbers and/or quality take years to put right.

France

2

Christophe Roturier

*Association de Coordination Technique Agricole,
149 rue de Bercy, 75595 Paris cedex 12, France*

Introduction

French agriculture has expanded to an unprecedented level since 1945. If it was the task of farmers at that time to bolster the country's food security, then today this objective has been achieved and exceeded, because France has become the second largest world exporter of agricultural and agri-food products.

The meaning of 'food security' has changed over the past 50 years. Former criteria of volume and quantity have been to some extent replaced by notions of safety of the food supply. However, current economic statistics should not distract attention from the challenges looming in the future: economic globalization, trends in the Common Agricultural Policy (CAP), more exacting demands from consumers with respect to quality, environmental and regional development concerns, and rising world demand for food.

This chapter describes the applied agricultural research and development system created in France during the past 50 years or so. This unique system, in which farmers occupy an important place, is responsible to a significant degree for the enormous agricultural expansion that has taken place. The system is described, and its strengths and weaknesses are assessed. Tropical research, mainly in the hands of CIRAD (Centre de Coopération Internationale en Recherche Agronomique pour le Développement) and ORSTOM (Office de Recherche Scientifique et Technique Outremer) will not be dealt with here. Similarly, the links with European research will not be examined.

©CAB INTERNATIONAL 1998. *Investment Strategies for Agriculture and
Natural Resources* (ed. G.J. Persley)

Historical Background to Present System

The system described below is the result of a long process, the origins of which are found in the 19th century, and which continues to evolve, as will be discussed later in the chapter. The history and the present configuration of the National Agricultural Development Association (ANDA) (see below) provide material that assists reflection on the possible evolution of other agricultural systems in the world. A study of the economic, social and political context that led to the emergence and structuring of farmers' associations in France holds many lessons for those concerned with rural organization and its impact on development mechanisms.

The following material, which in no way gives the full historical picture, focuses mainly on the modern period and emphasizes certain key events crucial to an understanding of developments in the French agricultural movement. Mention should be made first of the legislation of 1884 that recognized trade associations (*syndicats professionnels*) and permitted groups of farmers to set up the first multipurpose unions or societies to deal with their various concerns, from protection of their interests (steps against fertilizer fraud, for instance) to availability of technical support. The economic context of the time was marked by the industrialization process and widespread rural exodus. Concurrently, the labour movement was gathering force, enveloping the France of the Third Republic in major political conflicts. As a result of these conflicts, the agriculture sector split into two camps: on one hand the Société des Agriculteurs de France (SAF), representing the landowners for the most part, and on the other the distinctly republican Société Nationale d'Encouragement à l'Agriculture (SNEA), created at the instigation of Gambetta. Both organizations existed prior to the law of 1884. This dichotomy was a feature of the French agricultural landscape for a long time, and vestiges of it remain today. Other legislation followed, rounding out that of 1884; in particular, the 1901 legislation governing associations, a statute that greatly facilitated organization of the cooperative and mutualist movements.

The early part of the 20th century saw the appearance of production associations, known as 'specialized associations'. For instance: the Confédération Générale des Viticulteurs (wine growers) in 1907; the Confédération Générale de Planteurs de Betteraves (beet growers) in 1921; and the Association Générale des Producteurs de Blé (wheat growers) in 1924. The managers of organizations like these are often individuals with valuable technical qualifications. After World War II, there were rapid advances in technical areas. This new knowledge was disseminated by the state through the Agricultural Services Departments, and also by private enterprise, sector vocational organizations and small groups of farmers intent on keeping up with new techniques and associated with the Centres des Études Techniques Agricoles (CETAs). In fact, it was in these groups that the movement by farmers to take charge of their own development originated.

Subsequently, during the 1960s, there was a gradual transfer of state prerogatives in the development arena to the agricultural profession. An initial step in 1959 created links between the agricultural branches of the civil service and the corps of technicians attached to farmers' organizations; spread the technical progress within the context of the framework legislation of 1960 and 1962, which prescribed a structural policy and a mode of economic organization. The Fonds National de Vulgarisation (extension) et du Progrès Agricole, the precursor of the National Agricultural Development Association (FNDA), was also created at this time.

Beginning in the 1950s, the specialized associations developed technical groups, the Instituts et Centres Techniques Agricoles (ICTAs), to respond to the needs of their particular individual production branches or subsectors. A new phase was then launched in 1966 by a statutory order substituting the principle of 'agricultural development' for that of 'agricultural extension'. Farmers and their organizations thus became partners and actors in their own development, components in a dynamic linking research, training, and extension. This was also the point at which FNDA and ANDA were created.

The new system brought into being was founded on four principles: responsibility for the agricultural profession, decentralization of actions in the sector (coordinated by the chambers of agriculture in the *départements*), the profession itself as the source of financing and partnership between regions and production subsectors. The divestiture by the state which occurred in France at this period was also to be observed in many other countries, but in often very different political and socioeconomic contexts. For one thing, French agriculture at the time was in full expansion and in the throes of modernization; for another, this divestiture took place against the background of a strong impetus towards responsibility and independence on the part of farmers.

However, despite this generally favourable context, it must be pointed out that the reassignment of civil service personnel from the agricultural services departments to the chambers of agriculture sometimes created major personnel problems, attributable to the difficulties involved in the changeover from a public sector system of management to a system under the direct authority of managers belonging to the agricultural profession. This is a factor that should be taken into consideration in the current discussion process in some countries on withdrawal of the public authorities from the field of agricultural extension.

Agricultural R&D Structure and Organization

The system is treated as having four levels: farmers, development agencies, institutes engaged in applied research, and those involved, both public and private, in upstream research. In practice, the boundaries separating the different groups and their roles are not entirely clear. Before each of these levels is discussed in greater detail, it should be noted that ongoing technical exchanges take place between the groups, making the system effectively interactive. It is no longer the kind of system where the findings of upstream research are interpreted and disseminated to farmers by intermediaries. Analysis of farmer demand, its translation into research topics and recognition of farmer inventiveness in the face of new constraints have become essential stages in steering agricultural research in the right directions. Conversely, the dialogue with upstream research and the conduct of applied research leading to practical application, based on upstream findings, are indispensable to effective technology transfer.

The foundation of the entire structure is the farmers themselves, whether considered as individuals or as a community that comes together in Farming Study and Development Groups (GEDAs), Centres for Agricultural Technical Studies (CETAs) or Farming Equipment Cooperatives (CUMAs). Development at the local level is centred essentially around the chambers of agriculture, cooperatives and business enterprises.

The chambers of agriculture in the *départements* (administrative districts) are statutory bodies in which the majority of seats are held by Farmers' Unions (*syndicats agricoles*), meaning that they are controlled by the farmers. Although the chambers coordinate agricultural development (i.e. including extension) activities in the *départements*, their mandate is actually broader since they also intervene in tourism, local economic affairs, land questions and other important matters. However, their role in agricultural development is a major one owing to their presence throughout the country and their involvement in promoting GEDAs and, on occasion, CETAs. These grassroots groups have played, and still play, a primary role in France's agricultural development. They are set up on local initiative, often originating in the Farmers' Unions, and are powerful catalysts for the adoption of new techniques, given the community debate and exchanges of views they provoke. They also play an important part in bringing new responsible farmers into the picture.

The chambers of agriculture in the various *départements* making up one region are grouped into a Regional Chamber of Agriculture, which coordinates their activities through a Multi-year Agricultural Development Programme. Roughly 90% of all farmers in France are members of cooperatives in the various spheres of production. There are about 4000 such cooperatives. They provide their members with technical support and advisory services, which will vary in scope and complexity depending on their size. A similar role is played by the technical sales agents employed by firms selling to and buying from farmers.

Even if other groups make a contribution, *applied research* is mainly in the hands of the agricultural technical institutes and centres (ICTAs), which are essentially farmers' organizations (presided over by farmers), the first one having been created in the 1950s on the initiative of the specialized producer associations concerned. France has 15 ICTAs in all, each focused on a particular product or group of products (cattle, fruits and vegetables, grains, hogs, maize, etc.), unlike development organizations, which take a territorial approach. These 15 organizations are located at points throughout France employing some 1400 individuals, including 1000 professionals with advanced university or technician-level qualifications. They are grouped under the umbrella of the Association for Agricultural Technical Coordination (ACTA), which represents them before farmers and official authorities and organizes the mechanics of inter-ICTA operations involving topics (e.g. quality, the environment) of common interest to all production subsectors. They play a key role in technology transfer in agriculture, since they operate at the interface between the needs of farmers, the agri-supply and agri-food industries, society at large and upstream research. (More detailed information on these organizations is given later.)

Upstream research is conducted in the private and public spheres. Private research is carried out in the agricultural sector by agri-supply companies such as: animal and plant breeding, animal nutrition, agricultural machinery, fertilizers and plant health products. Public sector agricultural research in France is characterized by the importance of specific public agencies distinct from universities or other higher education institutions. For instance, the National Agricultural Research Institute (INRA), with 8500 staff members, the National Centre for Agricultural Machinery, Agricultural Engineering, and Water and Forest Resources (CEMAGREF), with 1000 staff, and the National Centre for Veterinary and Food Studies (CNEVA), with 600 staff, account for the major part of the public research effort in the sector. These agencies are well known internationally.

Dialogue between the different levels of this 'chain of progress' takes place in various ways: between individuals, between teams, between upstream and applied research groups, and also more formally via cross-participation in the scientific advisory boards of research organizations. For instance, the chairpersons of the scientific advisory boards of the ICTAs often come from INRA. Exchanges between applied R&D groups also take place informally between individuals, but they also occur in organized fashion as part of the consultation process initiated by the chambers of agriculture. In addition, some ICTAs have set up regional commissions or subsector committees whose mandate is to provide an upward path for information on the needs of farmers and their advisers. Dialogue between development organizations and farmers takes place mainly within the grassroots groups identified earlier.

There are also numerous initiatives aimed at facilitating exchanges between levels: framework agreements between INRA and ACTA, between INRA and the National Federation of Farmers Unions (FNSEA), between particular ICTAs and particular chambers of agriculture, and so on. Furthermore, it should not be forgotten that there are farm organization leaders present who manage the first three levels mentioned earlier: individual farmers and their local groups, the development (technical) level and the applied research level. The dialogue between these groups is considerably facilitated by the fact that the same farmer can chair an organization with a territorial focus (a Chamber of Agriculture), another with an economic focus (a co-operative), and a third with a national focus (e.g. an ICTA).

Nevertheless, it would be advisable to strengthen the consultation between research groups (public, farmers' organizations and industrial), development groups and the economic players involved. This would reinforce the continuous connection which ought to stretch from upstream research all the way down to the end-user. In other words, this is nothing less than the problem underlying the transfer of knowledge and technology: a problem we will return to shortly.

Financing the system

Finally, in concluding this general description of the French agricultural R&D system, something needs to be said regarding its financing, especially the financing of the organizations run by farmers themselves. The basis of the system is the mutualization of resources generated by quasi-taxes levied on agricultural products. The proceeds of these charges are channelled into the National Agricultural Development Fund (FNDA), which is administered by the National Agricultural Development Association (ANDA) and had a value of roughly Fr 770 million (US$128 million in 1996). It was initially funded almost exclusively by taxes on grains, a gesture of solidarity from the grain subsector towards other subsectors at a time of favourable grain prices under the Common Agricultural Policy. Moreover, the fact that the grain subsector has been organized since 1936, with establishment of the National Intervocational Grain Board (ONIC), guaranteed payments to producers and so laid the groundwork for the intro-duction of quasi-taxes.

Today, the system has expanded considerably, with the grain sector now accounting for less than 20% of the FNDA. The principle followed since 1996 has been that eventually every production sub-sector will contribute to the FNDA in accordance with its importance in the agricultural economy. In addition, a flat-rate farm tax of Fr 500 (US$83) has been introduced. This new system was instituted after a lengthy period of debate, and was finally judged preferable to a farm turnover tax. However, the FNDA's Fr 770 million (US$128 million) represents only a portion of the resources mobilized for applied research and agricultural development, which is estimated at roughly Fr 3 billion (US$500 million) including agribusiness.

The differential comes from local taxes raised by the chambers of agriculture, tax revenues allocated directly to certain ICTAs, miscel-laneous charges, service fees, and study or research contracts with the private sector (business firms) or the public sector (ministries, Euro-pean Union, product boards, local governments). There is also a trend for the contribution from *départements* and regions to expand under the decentralization policy introduced in France in the 1980s; although this can make it more difficult to 'read' some actions (Who is actually running them? With what aims in mind?).

ANDA: a Joint State/Farmer Agricultural Organization

ANDA, a joint state/farmer organization, which occupies a central place in the French R&D apparatus, is the expression of a unique farmers' development management formula, which is worth exam-ining in some detail.

Established in 1966, ANDA brings together agricultural organ-izations and the public authorities (Anon., 1993a). Its governing body consists of two groups, each with 11 members, one representing farmers' organizations (unions, chambers of agriculture, and cooperative, mutual and farm credit entities), and the other representing the state (Minister of Agriculture, Minister of the Economy and Finance). ANDA's man-date is to decide the major lines of development activity and to co-ordinate the programmes that it sponsors but does not itself carry out.

The Fr 770 million in FNDA, which ANDA administers, comes from the different subsectors of agriculture and from a flat-rate farm tax (representing just over 27% of ANDA's revenue; other main con-tributors are meat, 21%; grains, 18.5%; milk, 12%; and wines, 9.4%). These funds are distributed among the different applied R&D insti-tutions, each of which signs an agreement with ANDA. ANDA can also ensure that the programmes it sponsors are consistent with one another.

It is important to note that ANDA's commitment to a project often acts as a catalyst for the mobilization of funds from other sources (local governments, for example). In such cases, ANDA support serves as a measure of guarantee of the quality of the project and its consistency and coherence with other initiatives. Finally, as FNDA promotes the sharing of resources within and between the various subsectors of French agriculture, i.e. some form of collective solidarity, ANDA allocates some resources to projects that involve international cooperation and link French farmers and technicians with farmers in other countries, particularly in the developing nations.

Agricultural Technical Institutes and Centres

The role the ICTAs play at the interface between upstream research and agricultural development has already been mentioned. They occupy a special niche among the farmers' organizations, in the sense that they enable farmers to keep in touch with the world of research. These unique organizations are worth looking at in more detail. They were created on the initiative of the special associations, which are organizations grouped under the umbrella of the National Federation of Farmers' Unions (FNSEA). In order to promote dialogue among the various special associations and forestall possible conflicts of interest among the production subsectors, the Committee for Coordination of the Special Associations (CCAS) was formed. It is chaired by one of the vice presidents of FNSEA. To ensure coherence throughout the system, the CCAS chairman is also ex officio chairman of ACTA, the body created in 1956 to group the ICTAs into a federated whole.

The 15 ICTAs grouped within ACTA have articles of incorporation that vary considerably, they are of different sizes, and are active in different fields. They can be strictly farm production-oriented or include several categories of actors in the marketing channel. In the second instance, in addition to producers, the ICTA will include other players in the particular subsector (traders, distributors, transport agents) which is the case, for example, with the following: the Inter-vocational Technical Centre for Fruits and Vegetables (CTIFL), the Technical Institute for Vineyards and Winemaking (ITV), the Technical Institute for Hog Farming (ITP), and the Intervocational Technical Institute for Perfume, Medicinal and Aromatic Plants (ITEIPMAI). Obviously, the work done by the institutes with an intervocational focus goes beyond the strict field of agricultural production. CTIFL, for instance, might work on problems surrounding the packing of fruits and vegetables, or even organize displays on retailers' premises. ITV, in the course of its activities concerning winemaking, may work on questions of microbiology or processing hygiene.

Depending on the economic importance of the production sub-sectors on which they are focused, the ICTAs can vary significantly in size. For instance, the Technical Institute for Grains and Fodder (ITCF), which is also responsible for grain, legumes and industrial potatoes, employs 330 persons, whereas the Livestock Institute, which covers cattle, sheep and goats, employs 260. At the other end of the scale is the Technical Institute for Flax Fibre, which has only ten employees. Some of the ICTAs are concerned with a single product, for example, the General Association of Maize Growers (AGPM), or the Technical Institute for Hog Farming (ITP). Others may deal with a great many, for example, the Technical Institute for Poultry (ITAVI) works on 20 species, whereas the Intervocational Technical Centre for Fruits and Vegetables (CTIFL) works on literally hundreds of products.

So there is no one single model. Each single-product or inter-vocational group has developed a technical tool, the present character-istics of which are the result of changes over a 40-year period. This process of gradual change is still under way if one is to judge by a number of recent mergers, or even dismantlings, of ICTAs. For example, over the last few years the Institute for Rural Management and Eco-nomics (IGER) was replaced by the National Council on Rural Economic Centres. The National Intervocational Centre for Horticulture (CNIH) was replaced in 1997 by ASTREDHOR, which coordinates a network of horticultural research stations. The Technical Institute for Sheep and Goat Breeding (ITOVIC) and the Technical Institute for Cattle Breeding (ITEB) were merged in 1991 to form the Livestock Institute.

Such changes are evidence of the real influence of farmers them-selves in the life of their organizations. This influence is also clearly seen in the extent of their financial involvement in the ICTAs, which draw an estimated 60% of their resources from the agricultural com-munity. This estimate includes quasi-taxes channelled through ANDA, those levied directly on the ICTAs and contributions paid by farmers. Public funds (state, European Union, etc.) account for approximately 13% of the resource flow, originating more often than not in the context of research agreements (incentive credits).

ACTA plays a special role where public funds are concerned, since it works closely with the Ministries of Agriculture and Research in administering a special national budget appropriation for R&D. These funds are intended to facilitate joint programmes between the ICTAs and upstream research organizations. Over the 1993–1996 period, cor-responding with ACTA's Third Scientific and Technical Framework Plan, a total of Fr 60 million (US$10.1 million) was contributed to 56 research projects, with an estimated total cost of Fr 178 million (US$29.9 million).

Discussions are currently under way with the public authorities and ANDA on how to increase these resources. The aim is to improve technology transfer in agriculture, especially in spheres that cannot be handled by a single ICTA focused on a single production subsector. This is the kind of situation observed with environmental or regional development problems, for instance.

Farmers' Impact on Research Choices

Farmers can play either a direct or an indirect role in guiding research. It is direct within the organizations they control or with which they are associated, and indirect when exercised through ICTAs. In the main, the ICTAs are the research organizations controlled by farmers, who exert their influence in these at management level (in their Boards) and through the preparation of research programmes. As a result, ITCF has established regional committees with members who regularly bring together the farmers from a particular region so that they can present their problems and expectations *vis-à-vis* the institute. Their needs are then coordinated and discussed in national committees, according to crop category (e.g. wheat, grain legumes). Priorities are then set and submitted to the Institute's Scientific Council, which expresses its views and makes recommendations on the proposals. The Board then decides on the implementation of the various programmes.

It should be remembered that the Scientific Councils of the ICTAs, which are often chaired by INRA scientists, provide a valuable forum for dialogue between farmers and those working in government-funded research, who are strongly represented there. They provide an opportunity for the ICTAs to validate their methodologies and place their activities in a wider-ranging scientific context. Conversely, they enable government-funded research agencies to be better informed with regard to the needs of farmers; in this way, they can be regarded as an indirect channel for influencing the course of government research.

Without going into the various methods the different ICTAs have devised for analysing demand, mention might be made of the Stock Raising Institute's committees for the various subsectors (e.g. beef cattle, sheep and goats). Their proposals are examined by the Scientific Council before being submitted to the Board, in accordance with a system similar to that of ITCF. Similarly, CTIFL has established committees for fruit, vegetables and regional experimental stations. It should be noted that these stations are managed by CTIFL engineers, even though the stations themselves have been created and staffed by local authorities. CTIFL therefore plays an important role in the supervision of the scientific research conducted in these regional stations by providing a perspective at national level.

In addition, it must not be forgotten that upstream of the ICTAs farmers have a direct influence on research by managing FNDA jointly with the Government; this agency is an important source of financing for the ICTAs, as we have already seen. Consequently, farmers can have an impact on overall policy making as a result of the links between ANDA and the ICTAs. Finally, farmers can also directly express their wishes at the management level of public research institutions; for example, FNSEA, Centre Nationale des Jeunes Agriculteurs (CNJA), Assemblée Permanente des Chambres d'Agriculture (APCA) and Confédération National de la Mutualité de la Cooperation et du Crédit Agricole (CNMCCA) are members of the INRA Board. Nevertheless, it must be acknowledged that farmers can directly influence public research agencies only to a limited degree. In this respect, it would be helpful if, in general, the beneficiaries of this research (i.e. producers, processors and consumers) could be more closely associated with the tasks of defining strategies, selecting projects, and assessing performance.

On the other hand, the role played indirectly in influencing research through the ICTAs is considerable, although difficult to assess. In fact, it is usually the ICTAs that deal with research issues arising in the field, operating in accordance with the procedures described above. In such cases, they provide an interface with upstream research. Consequently, the dialogue and mutual interests existing between teams engaged in upstream research on the one hand and applied research on the other can lead to the emergence of joint programmes. On occasion, special financial support (in the form of funding for a thesis or research financing from ACTA's funds) will induce a research laboratory to investigate an area for which it feels no direct responsibility, because it is tangential to its normal activities or too close to applied research to be appreciated by specialized publications.

The importance of the Scientific Councils in the various agencies is therefore quite evident. Although government research is strongly represented in the ICTAs, the latter also participate – albeit to a lesser degree – in the activities of agencies concerned with upstream research. While ACTA does participate in many of the ICTA Scientific Councils, it is also represented on the Council of ACTIA (Association de Coordination Technique pour l'Industrie Agro-Alimentaire, which coordinates the activities of the Technical Centres and the Centres de Transfert de l'Industrie Agro-Alimentaire). The objective is, naturally, to promote the development of projects covering the whole of a production subsector, from production to processing or distribution. Conversely, ACTIA is represented on ACTA's Council for Scientific and Technical Orientation. Consequently, ICTA and ACTA are deeply involved in a large number of scientific networks, both formal (like the Scientific Councils) and informal. These make it possible for the participants to conduct an ongoing dialogue and influence one another.

Such exchanges are also greatly facilitated by the fact that the experts in the various agencies specializing in upstream research, applied research, development and industry are very often products of the same training ground; namely, the agricultural Grandes Écoles, to which one is admitted by competitive examination after 2 years of preparatory courses. This form of training is very different from the university systems found in most other countries. It produces engineers who may or may not continue their education by obtaining doctorates, and who may ultimately engage in very different occupations, but who have initially attended the same schools and studied under the same professors.

As a result, there is now no longer (as perhaps was the case formerly) a presumed hierarchy among individuals, based on the position they occupy in the R&D continuum. Nevertheless, even if INRA, the ICTAs, the chambers of agriculture and agribusiness enterprises hire their staff from the same schools, there is cause to regret that there is still little mobility among these different links in the chain and that, until quite recently, it has been in one direction only (from applied research or development to industry).

The Agricultural R&D System in France

The importance was emphasized earlier of the historical background in shaping the current French system, which has been built up gradually and continues to develop today, as demonstrated by the recent reforms in ANDA. Nevertheless, it is a complex system, with many parties involved. However, it continues to generate new forms of agricultural and agribusiness organizations, such as the Regional Centres for Innovation and Technology Transfer and agricultural towns (agropôles). Although such profusion could be regarded as a symptom of dynamism, and inspire competitiveness, it may sometimes amount to nothing more than a mere dispersion of efforts, or indicate that resources are being wasted.

Moreover, the particular way in which the ICTAs have developed historically, and which has caused them to specialize in individual production subsectors, may certainly have made it possible to deal effectively with difficulties affecting particular products. However, it has sometimes led to the creation of small organizations, posing problems in terms of both scientific efficiency and management of human resources and support services (e.g. biometrics and computer systems). Some professions have responded by passing on the study of technical problems affecting their production subsector to larger agencies; for example, this has been the case with grain legumes and industrial potatoes, which are being studied by ITCF on the basis of agreements among farmers who have arranged to subcontract these technical studies.

Another feature of the French system is that the farmers generally have little control over decisions regarding the direction upstream research in the public sector is to take. Some people regard this as a symptom of the freedom essential for inspiring creativity among researchers; others consider it as merely allowing research to drift away from the goals originally set. These contrasting viewpoints also pose the question of how research is to be evaluated; i.e. by whom, how, and for whose benefit is it performed? The state, for its part, identified in 1996 a number of priorities for national research and these priorities include agriprocessing and the environment. In order to induce public-sector researchers to direct their efforts towards these priorities, Conseil Interministeriel de la Recherche Scientifique et Technique (CIRST) has also decided that a major portion of the funding of laboratories (20%) should eventually come from programmes administered by research agencies.

Finally, it should be noted that until recently the young staff members trained by agencies specializing in applied research or development were very often recruited into industry once they had acquired a few years of professional experience. More recently these losses of staff trained by the farmer agencies have been less prevalent, but they can, nevertheless, have certain positive effects; i.e. by ensuring that future industrial managers will be more sensitive to farmers' problems and expectations. Moreover, when this sort of mobility occurs at a later stage in careers, it can result in the development of valuable networks of contacts.

Nevertheless, agricultural technology transfer would certainly become more efficient if there was more mobility between the various units active in R&D, and the movements were multidirectional. Such mobility promotes the exchange of ideas, the convergence of opinions and the adoption of new methods. With respect to public-sector research, this issue once again raises the question of evaluations and career-management as applied to researchers.

The French system must, of course, be given credit for the considerable overall research capacity it offers through the combined efforts of all those involved, whether they are in the public or private sectors, or farmers' organizations. With regard to private-sector research, an undeniable contribution has been made by large concerns such as Rhône-Poulenc and Limagrain. As for public-sector research, the concentration of resources into a few clearly identified and internationally recognized institutions, such as INRA and CEMAGREF, results in greater efficiency than if they were dispersed among a large number of operators, and also makes it easier to identify the appropriate contact persons and units.

The farmers' organizations specializing in applied R&D are the most original feature of the French system. One of their main assets is that they are in direct contact with farmers, and thus know their needs. Furthermore, because they are at the interface between upstream research and the farmers, this group of experts can express in scientific terms the questions that arise in the field and, conversely, put research findings to practical use. Because of them, the dialogue between upstream research and the farmers themselves can be put on a scientific footing.

Finally, the control by farmers of organizations specializing in R&D means that when financial resources are less abundant, they themselves can select among the sometimes conflicting interests of the various production subsectors or regions. The recent reforms in FNDA provide a good illustration of this, as does the decision to give priority to supporting applied research by maintaining the funding of the ICTAs. In contrast, when farmers in leadership positions consider that an institution is no longer playing a relevant role, they can – purely and simply – eliminate it (as happened with the Technical Institute for Apiculture in 1993 and the National Interprofessional Horticulture Centre in 1996), even though they do run the risk of dismantling research mechanisms that might be needed a few years later.

There is overall agreement that the French agricultural R&D system provided sound responses to the demands for increased productivity in the postwar period. Today, there is still room for improvement, since more attention could be given to the needs of processors, consumers and society in general, and the transfer to users of the applications of research findings (in the form of know-how and methods) could be speeded up. The problem of such transfers is quite a common one in France, as witnessed by the recent efforts of the ministry responsible for research to establish Technological Resource Centres, whose function would be to identify the needs of small and medium-sized enterprises and facilitate their access to research laboratories that could solve their problems.

Europe as a whole is also giving close attention to this issue, as is shown by the current discussions of these matters in the European Commission and the member countries, in preparation for the Fifth Framework Programme for R&D. In addition, the new Agricultural Orientation Law (now in preparation) is intended to set the major objectives for French agriculture over the coming 10–15 years, thus defining a framework for R&D.

Conclusion

The French agricultural R&D system is complex and difficult for the outsider to decipher. Not only is it the heir to a series of historical circumstances peculiar to France, but it was also built up in an economic context (the 1960s) and a regulatory context (the former Common Agricultural Policy) favourable to agriculture. It would therefore be risky to transpose it as it stands to other situations. However, it does provide a number of key lessons for those engaged in creating such systems today:

- The uniqueness of the French system lies in the distribution of the different links in the R&D chain between the state and the farmers: the state is responsible for basic and mission-oriented research, whereas both applied research and development are guided and directed by farmers. In some countries today there appears to be a convergence between the policy of state withdrawal from certain R&D areas (extension, at least) and the need to consider demands from farmers. This appears to be an opportunity that should be seized so that farmers in these countries (and the private sector) can take over some of the areas in question. However, we have already seen just what an effective organization of farmers requires in terms of legislation, farmer experience, human resources training (for leaders, in particular), financing, etc. It is in these cases that study of the history of the French agricultural movement can prove a valuable source of instruction on how to set up representative and structured farmer organizations.
- It should be possible, using the experience of France, to imagine less complex systems incorporating fewer structures. Consultation and collaboration between these structures and the State would then be easier, especially with regard to the determination of areas of competence and procedures for reassigning responsibilities. Formulation and implementation of a national agricultural policy would also be simplified. However, it must be noted that careful human resource management is important when reassignments of jurisdictions and responsibilities are accompanied by personnel transfers. This is not an issue to be treated lightly, since properly qualified and enthusiastic staff are crucial to the success of this type of undertaking.
- Finally, in circumstances where farmers' resources are scarce, it would appear wise to concentrate them in applied research, which requires fewer resources than development, allows better preparations to be made for the future and facilitates relations with the world of research in general.

By way of example, it is worth noting that in France the budget appropriation for the ICTAs was roughly Fr 890 million (US$148 million in 1996), whereas the full cost of the applied R&D system is estimated at Fr 3 billion (US$500 million).

In all these scenarios, close attention needs to be paid to the quality of the links between upstream research, applied research, basic and continuing training and extension to ensure the transfer of new knowledge and techniques.

Germany

3

Alois Basler

Institute of Agricultural Market Research, Federal Agricultural Research Centre, Braunschweig-Volkenrode, Germany

Introduction

An analysis of the agricultural research system in Germany is presented in this chapter. It describes the global characteristics of institutional structures, highlights strong and weak points, outlines constraints that have occurred in recent years, particularly those that resulted from German reunification, and prospects for the future. The perspectives emphasize the potential contributions of German agricultural research in an overall global international research system.

The chapter summarizes many different issues that can only be dealt with here by highlighting the main issues. Details, for example, on the old and new research structure in Eastern Germany, or budget and funding figures, can be obtained from other published papers. It must be noted that the restructuring process is still under way and the structure that will emerge at the end of the adjustment process will depend on many factors.

Federalism: Impact on Research and Training

Basic decision favouring subsidiarity

The organizational structure of agricultural research in Germany is determined by the political and administrative structure of the country. After World War II, decision-makers opted in favour of a decentralized system of political power and decision making. The country is subdivided into (now 16) states (Länder), each with political autonomy in decision making by their own parliaments, governments and ministries on key areas of economic and social life. The state governments were given the exclusive responsibility for research and training at all levels (elementary schools, high schools, universities). Each regional government assumes these responsibilities in its own fashion. Decentralization of decision making has its corollary in the autonomy of income generation by taxation. Regional governments have the right of tax imposition and collection in certain well-defined areas.

According to that fundamental decision on sharing the power and responsibilities between the federal and the regional level, research in general and consequently agricultural research and training are carried out first and foremost under the supervision of regional ministries. Planning of research and training and supervising activities are exclusively under the rule of the Ministries of Scientific Research, Education and Training in different regions. They provide necessary funds which come exclusively from their own tax revenue. This phenomenon is referred to as *regional research*. Federal ministerial services only participate in financing part of the infrastructure of universities, but have no influence and no responsibility for research and training, priority setting, organization and control. Research in Germany is considered to be first and foremost the task of universities. This is in contrast with other European countries, where centrally organized research is the centrepiece of the research landscape and constitutes the base of technical, economic, cultural and social progress.

It may appear that such a division of responsibilities between different regional entities would entail a large heterogeneity of training programmes and render mutual recognition of university diplomas of different regions more difficult. In fact, this has been and still is a crucial issue of the federal system. In order to cope with the problem of coordination, regional ministries for scientific research initiated and encouraged universities to establish the Scientific Council (SC) at the federal level, incorporating the regional administrations and representatives of universities. The SC has provided clear evidence of a high quality of work and was particularly active during the reunification process.

In spite of this fundamental decision, it is still possible to carry out public research activities at the federal level. If ministries at the federal level have defined needs for information to conduct their daily work and to carry out assigned responsibilities, the fundamental law gives them the opportunity to establish, organize and finance research under their direct responsibility. Research under their control, and paid for with funds from their own budget must, however, meet urgent needs related to their own decision making, and cannot go beyond that restriction. This general rule opened the way for establishing an agricultural research structure at the *federal* (*national*) level (called federal research in this chapter) which is placed under the rules and regulations of the Federal Ministry of Food, Agriculture and Forestry. It is worth noting that the Ministry of Agriculture is one of the few ministries that seized the opportunity to endow itself with a relatively large research infrastructure covering the main research and policy fields in agriculture. Other federal departments have small research departments, but generally they collaborate with research institutions of regional ministries if research inputs are needed.

Nevertheless, efforts have been made by other federal ministries intending to establish research groups, to provide them with technical information on a regular basis. These efforts resulted in the creation of so-called 'blue-list institutes', which are research units funded jointly by regional and federal ministries. In 1996, about 83 institutes of that type were in place, with approximately 1000 scientists involved in research programmes. They cover a wide range of research interests and partnerships. Research areas range from technology development in specific industrial branches to analyses of the overall development of the German economy. Until 1990, agricultural research was not represented in such institutes, except in the five macroeconomic research centres that deal with agricultural prices, markets and growth rates within their macroeconomic modelling. Research programmes of blue-list institutes are closely linked with and dependent on the information needs of sponsoring ministries and the respective industries which use research results or are affected by ministerial decisions originating in research institutes.

A fourth type of institution are those organized as *private but non-business-oriented* research centres. They receive substantial amounts of public funding. The Max Planck Research Centre and the Frauenhofer Association are the two most important institutions working in various research areas and dealing with a wide range of topics. However, with the exception of one plant breeding research institute, the key fields of agricultural research do not come under their mandate. They are, nevertheless, involved in several agricultural research concerns, and share activities and results with agricultural research institutions on the federal as well as on the regional level. The two centres mentioned had a total of about 4000 scientists in 1995.

Last but not least the *private business sector* deals with agriculture-related issues. There is a wide range of privately funded societies or institutions. Some are large and powerful commercial-oriented breeding, fertilizer and plant protection firms, others are smaller research groups that focus on environment-related topics.

Some of these groups are supported by philanthropic foundations. The consequence of this diverse picture of research interests and institutions is that the agricultural research landscape in Germany is widely scattered. A great number of large and small units are doing research on agricultural questions and issues. Someone who is looking for specific research results or who is searching for collaborators in a specific research field, programme or activity faces problems in finding and getting in touch with appropriate partners.

Research coordination at the federal level

The fundamental principles of sharing responsibilities in research remain unchanged. Regional sovereignty, leaving the major part of responsibilities in research and training to the states, was never questioned as a general rule. However, in the course of agricultural development, new research-related questions and problems arose that required decisions at the national level. Because of this, decision-makers agreed in 1965 to build up the Federal Ministry of Scientific Research, responsible for scientific promotion, nuclear research and space exploration. During the 1970s two ministries were set up dealing with research and education on the federal level. The Ministry of Research and Technology was in charge of:

- research coordination in general;
- promotion of research with respect to energy, biology and ecology;
- research on geology, procurement of raw materials and space exploration;
- information and production technology.

The Ministry of Education and Science was responsible for:

- principles and organization of the education system;
- improving professional skills;
- scientific policy.

In 1995, these ministries merged into the Ministry of Education, Science, Research and Technology (BMBF). Incorporating tasks of both ministries, it is now the leading national organization for research and training, which cannot be dealt with sufficiently by regional political entities. The Ministry is in charge of the so-called 'large research centres' (Großforschungseinrichtungen). Of these, the research area of the Society of Biotechnological Research (GBF) is closely related to agricultural research. Otherwise the federal agricultural research system continues to be organized and funded through the Ministry of Agriculture as before.

Evolving Agricultural Research Capacity and Structure

Structure and organization

Public research

Agricultural research in Germany has traditionally been a high-ranking research field. Successes in breeding, plant nutrition, plant protection, conservation of plant genetic resources, development of new farm equipment, animal breeding and animal nutrition have been achieved, thereby improving human nutrition and the living standards of the rural population. According to the guiding principle of subsidiarity, which has been and is still the centrepiece of state organization in Germany since World War II, agricultural research is carried out mainly by four institutions with differing styles and institutional affiliation:

- Agricultural faculties of the regionally based universities.
- Research centres and institutes of the regional ministries of agriculture.
- Federal research centres for food, agriculture, forestry and fisheries working under the responsibility and with budgetary backing of the Federal Ministry of Food, Agriculture and Forestry.
- Private sector business research.

First of all, agricultural research (as all other research activities) is one of the main tasks of universities. Before 1989, i.e. in the former West Germany, seven universities housed large agricultural faculties. All of them continue to cover the main scientific disciplines of soil science, plant production and protection, livestock development, agricultural engineering, food processing and economic and social sciences. The soil/plant complex, animal research and social sciences are the branches with the highest priority. Three universities have a faculty for veterinary medicine. In addition there are some other universities which are not engaged to a large extent in agricultural research, but which house some research groups working on issues related to agricultural topics, such as biology and geography.

In 1988 the total scientific staff involved in agricultural research and training in universities in Germany was about 3600. Agricultural research under the responsibility and management of universities is considered as an essential prerequisite for high quality research and for university training with high quality standards. Taking into consideration that university training leads, among other purposes, to a scientific career, research has been in the past, and will continue to be in the future, an essential component of the mandate of universities.

A second type of regional research is that conducted by the publicly funded regional research centres, placed under the supervision of regional ministries of agriculture. They are pursuing downstream, adaptive research. Their activities are targeted to serve and to satisfy needs of regional ministries, extension services, and technology transfer activities at the regional level. They maintain links with the regional faculties of agriculture. This type of research is relatively small, at about 10% of the federal research capacity referred to later.

The third type of research is that conducted by higher education – like training institutions (Fachhochschulen). Like universities, they are placed under the regional ministries and provide higher education. The research component of these institutes (each regional state has at least one such training centre) is very small, and there are no plans to expand their research function.

The Federal Research Centres (FRC) are responsible to, and controlled by, the Federal Ministry of Food, Agriculture and Forestry, with funds provided through the Ministry's budget. The mandate of these research centres is to conduct research in all fields of the Ministry's interest. Thus, centres are serving the needs of the decision-maker in all aspects of agricultural policy at the federal level. Before 1989, 13 research centres were engaged in different fields of research, with a total of 2701 employees, 900 of whom were scientists. In addition to these scientists, who had long-term employment contracts, federal-based (national) research centres have a scientific staff working on specific projects. They have a fixed-term collaboration contract (less than 5 years) for conducting a special research project, with supplementary funds provided by the Ministry or other public or private sponsors interested in a specific research topic.

The mission of these research centres is to meet the information needs of the Federal Ministry of Agriculture. In pursuing that task, research serves the following main objectives of the German agricultural policy as outlined in the Ministry's annual report of 1997:

- To improve the living conditions in the rural area, and to secure adequate participation of the agricultural population in the overall development of incomes and welfare.
- To provide the population with agricultural products of high quality at reasonable prices.
- To improve international agricultural trade relations and the world food situation.
- To secure and improve the state of natural resources, to maintain the biodiversity, and to improve animal welfare.

Private research
Public research is complemented by a wide range of private agricultural research as discussed below.

The chemical industries, which produce fertilizers, pesticides and other chemicals, have large research capacities at their disposal in order to develop and test new products and new pest management methods and procedures. These are being offered to users through their own distribution and extension systems. During the past decade the private fertilizer and pesticide industries have dedicated more and more attention to new research and development activities, including so-called 'smart' technologies of crop protection:

- Products with low toxicity.
- Better and safer product packaging systems.
- Products incorporating ingredients without dust formation during their use.
- Microencapsulation of ingredients.
- Protection of active ingredients from outside contact.
- Safer use of insecticide sprays.

The private German insecticide industry is greatly concerned about, and working towards, the development of less harmful crop protection products and procedures and safer biological control methods. This new emphasis and direction is mainly due to stronger public regulatory measures and a greater awareness of product users about chemical residues in products and concerns for their own health in using these products.

Private breeders are developing new product lines and new plants in their own laboratories and in their own test fields. The most important crops the private researchers are working with are wheat, coarse grains, oilseeds, sugarbeets, potatoes and some species of tree plants. In Germany most plant breeding is done by private breeding firms. They have their own seed multiplication and distribution systems that are, nevertheless, submitted to strong public regulations determining procedures of seed multiplication and trade. The main public responsibility in this field is to maintain and enforce appropriate laws and guidelines respecting new varieties and to operate a gene bank to conserve all genetic material, including new lines. The system of task-sharing between private and public research is approximately the same for animal breeding. Breeding cows for milk and meat production is mainly conducted by private enterprise associated with the German Breeder Association, which disseminates breeding results and sells breeding material. This task-sharing between public and private research is the result of an agreement between the Secretary of State of Agriculture and representatives of the private breeding business, reached in the early 1950s, making breeding a mainly private sector responsibility.

Another important area of private research covers all types of pharmaceutical products for human health and for the cosmetics industry. The private sector has a key role in the development of such products. Public research in these areas controls the composition of products and their suitability for use by the public.

Engineering and the agricultural machinery industry continue to improve technical performance of their products, using their own research and testing facilities. Promotion of new techniques and products and the transfer of technology is done mainly by private firms. Public-funded research has only a complementary role, and its capacity in these areas is relatively small.

Private business interests are strong in genetic engineering as a special branch of biotechnology. One would expect that research in genetic engineering would attract huge amounts of private funds and develop into a mainly private research domain. However, until the mid-1990s restrictive laws and regulatory requirements restrained to some extent such a rapid expansion of R&D activities in this area. Private enterprise invested outside the country, and scientists who specialized in this field looked for jobs in other countries and regions, resulting in a brain-drain of specialists. Public research and public-funded private, but non-profit-oriented, research institutions nevertheless did some useful work within the defined limits. However, significant changes are under way. The government asked the recently established Council of Research, Technology and Innovation to work on methods of making better use of biotechnology in Germany. The Council presented a comprehensive report early in 1997 which underlined the importance of this matter and elaborated some lines of programmes and measures to be undertaken by public and private decision-makers and R&D agencies. Taking into account the budgetary constraints of public institutions which have a lasting effect on research funding, the initiative can be considered as a green light for future private sector R&D activities in the field of genetic engineering in Germany.

Priority setting

Taking into consideration the extremely scattered research landscape in Germany, setting and coordinating priorities in agricultural research within the global national agricultural research system (NARS) is not an easy task. As far as university research is concerned priorities are set and determined by each faculty, institute or, to some extent, even by individual researchers themselves, provided that the regional ministries release funds.

Federal agricultural research
Priority setting of federal (national) agricultural research institutions follows a significantly different path. Research centres under the

authority of the Federal Ministry of Agriculture present every third year a proposed research plan for the next period indicating research areas. The planning procedures applied within centres or institutes may differ. In centres with large capacities, programmes are defined through participation of scientists at the institute or laboratory level, and then collated to complete the centre's plan. Relatively small centres usually apply a more top-down approach.

Private sector research
Management of business-oriented agricultural research is quite different. Most frequently, funding organizations and beneficiaries are identical. Research is done by the user. Objectives and priorities are exclusively determined and set by the market. The crucial problem and challenge public research is facing, i.e. meeting the needs of users, is automatically solved in business-oriented research which is demand driven.

Cooperation in research
Considering the two distinct institutional types of public agricultural research organizations (universities which are run by regional states and federal research under the auspices of the national Ministry of Agriculture, Food and Forestry), to what extent and through what mechanisms do they collaborate and coordinate their programmes, priorities and results to avoid duplication? Each research entity defines its programme and no official and formal programme coordination mechanism is in place. Federal agricultural research institutes work on behalf of the Ministry, and their research work is mainly downstream or adaptive. Universities are essentially free in the choice of their research topics, which have a stronger upstream orientation. Although the research topics are to some extent different, there are a number of cooperative activities between federal and regional agricultural research, allowing mutual use of knowledge, results and even resources:

- PhD students often work on research problems in federal centres, using their facilities.
- Researchers working in federal centres give lectures in universities through special agreements.
- Institutions from both sides carry out joint research projects.
- Institutions from both sides exchange, on an informal basis, work experience and results.

Conclusions
To sum up, there are, in the federal research system, close relationships between research institutions and the ministry that orders

research, and the ministry's influence on programmes is relatively strong. In contrast, agricultural faculties in universities run by regional states enjoy a larger degree of freedom in defining their research programmes, because regional ministries of education do not use research results immediately. Priority setting procedures have been and still are a major subject of research management discussion. German institutions, scientists and management advisers are involved in such discussion, particularly with respect to agricultural research in and for developing countries. Surprisingly, during adjustments in existing German research institutions, very few attempts were made to formalize the priority-setting process or to introduce indicators for quantifying priority criteria.

Linkages between research and users

Procedures and effectiveness of technology transfer

Agricultural research globally has two main clients or end-users: the farmers, concerned with best ways to produce agricultural products; and politicians, who take decisions on the whole juridical, economic and social framework of agricultural production as well as living conditions in rural areas. Research has to look at both sides. Taking into consideration the role of policy interventions in Germany and in the European Union with a broad, diversified and sometimes contradictory set of objectives during the past 40 years, the role of decision-makers as clients for research results has been increasing.

To respond to the information needs of *decision-makers* is primarily the task and the mandate of federal research institutes. As described above, their research programmes are established through a joint effort of research institutes and the respective services of the Federal Ministry of Agriculture. Thus, the user of results does have the institutionally anchored opportunity to influence research programmes and to push them in the desired direction in order to get the information needed.

The transfer of technologies to *farmers* is more complicated. The three lines of agricultural research in Germany are using different ways to diffuse their innovations. As far as private research is concerned, the utilization of results by farmers may be uncertain and constitutes a certain risk for the producer of research results, but the whole transfer goes through the market. Research results (e.g. new lines of products) are being supplied on the seed market at a given price. Breeders are selling their products and they have a vital interest in getting their seeds accepted and bought by farmers. A breeder will necessarily disappear from the market if product transfer fails for whatever reasons (e.g. bad quality of the product, inappropriate transfer structures and mechanisms). Suppliers get the feedback immediately. For that reason, private research will do its best to know the real needs of end-users. Considering the growing role of private research in various fields one may conclude that private research has been able to recognize needs and to organize the transfer process effectively.

The transfer process of public research results to users works differently. First of all, the farmers are not directly involved in or associated with the priority setting process, nor are their needs taken into account in programme design. There are no commercial constraints to require this, since the 'product' does not have to be sold. Producers may transmit their interests directly sometimes but mostly by intermediate organizations such as extension services, producer organizations and agricultural administration offices. These institutions constitute the main channel for bringing the user's interests to the research organization. To some extent, they diffuse results to farmers through extension operations. Thus, the effectiveness of technical transfer depends mainly on the working capacity and dynamics of these intermediate agencies and units, particularly on those whose first and unique mandate is extension.

In summary, one must recognize that there are various channels or spreading information and for transmitting research results and any other innovation on technological improvement of production, processing/storing and marketing of products. The channel with the highest frequency of utilization seems to be written information. This means that researchers and institutions must be interested not only in publishing their results in scientific journals, but also in bringing their message in a readable and understandable form to practitioners. Publication services of research institutions in Germany do not usually have this option.

Final results are presented to a broader public through scientific journals and information channels (mainly professional press services) of appropriate professional groups and associations, a way of dissemination that is also used to get messages to farmers. This last-mentioned channel is the normal and most frequently used route for spreading results among clients. In order to get the message accepted, it must be presented, highlighted and worded in a manner that differs significantly from scientific journals. The fashion of transmission must be attractive for potential clients. The methods of 'selling' research results to clients certainly need some improvement in Germany. Effective public relations services that can provide help are somewhat underdeveloped in public research centres and organizations in Germany. The question of how to get results transmitted to, and eventually applied by, potential clients had not been adequately addressed in the past. For some years substantial efforts have been under way to strengthen and improve the dissemination of research results.

Farmers' influence on the research agenda
A second and just as important information flow is the feedback research institutions receive from users of research results. Effectiveness of the whole knowledge exchange network, and the influence farmers have on the research agenda, depends to a significant extent on responses from farmers or other end-users. The top-down transfer structure provides some components of the answer to the question of how *farmers* are exerting such an influence. They can act by going through intermediate organizations (extension), which are not only in charge of transferring research results to users but also of feedback and ideas users wish to communicate to researchers and development centres. Extensionists are frequently and successfully involved in the functioning of the R&D system in Germany, particularly in providing valuable feedback to researchers. In cases where researchers are doing on-farm research, extensionists are often involved in the observations and problem assessment. However, because many research activities, particularly those done by PhD students, involve more and more laboratory analyses, this type of direct feedback is limited.

In the past, farmers directly influenced the research agenda by exerting influence on public-funded research programmes through the intervention of agricultural policy-makers. If, for example, the intensive exchange of views between farmers' associations and policy-makers results in strengthening specific programmes where research inputs are needed, the policy-makers may opt for making available supplementary research funds. Research institutions that intend to apply for those funds will focus their programmes on such issues. An example of the impact of this influence is the renewable resources programme. A few years ago, politicians in agriculture in the federal and regional levels considered renewable resources as a way to reduce the problems of subsidized agricultural overproduction. To get more information on procedures and costs, a broad research programme was launched incorporating a large range of institutes of federal and university research. Some institutions even changed their mandate declaring these issues as the focal point of research for the future. In those cases, the first initiative did not come from research but from the farmers.

Through informal activities, other clients have a certain influence on topics that research institutions are pursuing. This is particularly true with agribusiness. The informal collaboration structure between institutions and private firms allows firms to communicate to research institutions about the type of innovations that are marketable, and the ones on which they are working.

These non-farmer clients also include public and private institutions dealing with technical and financial cooperation. For some research institutions within universities, these development agencies provide an important contribution for financing operations in developing countries. In that way they contribute substantially to the qualification of young professionals specializing in research areas of particular relevance to developing countries.

International orientation

Links with other industrial countries

Postwar Germany took a strong world market-led approach, seeking out new markets and investment opportunities in foreign countries. Agricultural production was largely exempt from that open economy concept and was protected against international competition. Agricultural production was targeted mainly at satisfying national demand, and policy tended to be inward-looking. The main development target had been to restructure German agriculture (Germany was far from being self-sufficient in food) within a European context. The adjustment process was determined by the application of common and guaranteed prices for key products (cereals, beef, milk, sugar) in all EU Member States. The new environment for agricultural policy resulted in strongly subsidizing these production sectors, in a progressive centralization of agribusiness, and in a continued upward trend in production. By the end of 1970, Germany achieved self-sufficiency, and in the years following produced substantial surpluses of cereals, beef, milk and sugar.

In contrast with these protected markets, other sectors such as fruit and vegetables, were fully exposed to European and worldwide competition without price guarantees or producer subsidies. Fruit and vegetable production and processing in Germany thereby declined for some time. The processing industry recovered to some extent with a relatively important share of imported raw products.

Priorities and programme design in the federal agricultural research system more or less reflected the evolution of production. For almost 30 years, its main concern was the increase in production and yields and the improvement of market facilities for German agriculture. In the late 1970s agricultural policy and the corresponding research activities under federal governance took a more outward-looking orientation, emphasizing closer collaboration with other European countries (Burian, 1992). The Ministry supported and encouraged these efforts. The programme achieved remarkable results. Since the 1970s many contacts and agreements have been established with most Western European and North American countries. Even some Eastern European countries were (before 1989) involved in this bilateral collaboration. With regard to the duration and thematic scope of collaboration, results were rather modest, with only a few contacts leading to common research projects.

Universities followed a quite different path. During the 1950s university research had established close ties with the Anglophone research and scientific world, particularly with American universities. Undergraduate and postgraduate students spent time abroad, particularly in the USA, becoming familiar with new scientific methods and establishing long-lasting contacts. This exposure had an influence on their approach to teaching and research when they returned to Germany.

Research on and cooperation with developing countries

In contrast with most other European countries, Germany did not have traditional links and institutional connections with agricultural research in developing countries. Thus, any attempt to undertake activities in this field had to start from zero. Politicians and scientists shared the commitment and the conviction that Germany had to contribute to the overall effort to find solutions to agricultural development and food security problems in developing countries.

During the 1960s German institutions developed partnerships with research institutions in developing countries. Often these activities resulted from individual initiatives of German scientists who developed scientific interests in tropical agricultural development topics. They tried to find sponsorship for their projects. As a result, research themes were often defined by researchers in Germany and reflected research programmes pursued in their own laboratory. The objectives of sponsoring institutions were obviously taken into account. However, research topics and the objectives of cooperation with research organizations in developing countries were largely determined by scientists who came from Germany.

With the establishment of the Consultative Group on International Agricultural Research (CGIAR), research cooperation with developing countries as conceived by German university and research institutes entered a new phase. Germany underlined the commitment in favour of supporting the international agricultural research centres (IARCs) working under the CGIAR umbrella. Germany's contribution of DM 35 million (US$19.8 million) yearly is being channelled through the Ministry of Economic Cooperation which, however, does not have its own research structure for giving advice with respect to types of programmes to be sponsored.

In parallel with evolving CGIAR activities, German bilateral cooperation agencies expressed the need for research support to implement agricultural and rural development projects. Demand came particularly from: (i) 'Deutsche Gesellschaft für Technische Zusammenarbeit' (GTZ) for technical cooperation; and (ii) 'Kreditanstalt für Wiederaufbau' (KFW) for financial cooperation such as credit and loans. Many cooperative activities were established (Federal Ministry of Economic Cooperation, 1992). Cooperation between research institutes and executing agencies for bilateral cooperation was achieved in a number of ways:

- Research institutions were involved in analysing specific questions that arose from project implementation.
- Participation in preparing programmes and projects.
- Executing agencies used research staff for project implementation during a limited period.

To cope with these new challenges, agricultural faculties integrated developing country research components into their programmes. Four faculties declared this field as a priority task, and established a 'Centre for Tropical Agriculture' as a coordination unit between different departments and institutes. The federal agricultural research system is less involved in this area of research: only 6–8% of research capacities are dedicated to topics related to developing countries.

Scientists and administrators involved in such programmes constantly faced the problem of how to meet the demand of decision-makers for research results, with involvement from such a widely scattered research structure. Leading executives of the ministries involved, supported by some researchers, proposed the creation of a coordination bureau that would maintain ties with all types and levels of research institutions dealing with tropical agriculture. The initiative resulted in the establishment of the 'Council for Tropical and Subtropical Agricultural Research' (ATSAF) as a coordination unit for agricultural research in the tropics. ATSAF is an association of scientists from federal research institutes, universities and other research institutions working with special interest in the field of tropical and subtropical agricultural research and related fields. Its future is currently under review.

Funding mechanisms: a systematic review

To understand fully the uniqueness of the German agricultural research system and its potential role within an international context, an explanation of the different sources of agricultural research funding in Germany is necessary; these are listed below.

1. Long-term funding by public budgets through ministries:

- Federal Ministry of Agriculture financing federal research centres.

- State ministries (Länder) of agriculture funding regional research centres.
- Regional ministries of education funding agricultural faculties.
- Federal and regional ministries providing funds for so-called blue-list institutes that have at least two ministerial sponsors: one on a federal and the other on a regional level.

2. Special fixed-term funding which is provided to conduct research on specific topics determined by public agencies; funds may come from:

- Various federal ministries, mainly of agriculture, but also of education and research, environment and economic cooperation, for research projects in and about developing countries.
- Several regional ministries expressing needs for information that can only be covered by doing research on a specific theme.

3. German Association for Scientific Research (DFG), created in the 1950s as a public fund with the objective of promoting scientific research in Germany. Besides ministerial budgets for universities and research centres, DFG funds play a pivotal role in supporting young scientists doing a first or second degree or a PhD. The fund, administered by the Federal Ministry for Education and Research, is also open for specific agricultural or agriculture-related research programmes, even if they are conducted outside Germany. However, support is provided only for German scientists.

4. Private foundations supporting programmes proposed by any institution; one of the foundations having played an important role in agricultural research in the past is sponsored by Volkswagen.

5. Funds coming from outside and particularly from various EU budgets related to specific programmes or projects with a specified expected outcome.

6. Private business-oriented research.

Constraints and New Challenges for German Agricultural Research

The institutional structure of agricultural research in Germany has evolved over 45 years. During the 1950s and early 1960s, capacities expanded in a rather moderate way, owing to small public budgets. During the 1970s, the total research and higher education system entered a period of rapid growth in terms of number of new institutions and staff. Agricultural research followed this general trend. Furthermore, during that period, and specifically after the serious food shortage in the Sahel in 1973–1974, the R&D community dedicated more and more attention to agricultural and rural development and other problems in developing countries. This is often used as justification for expanding existing research capacities, and fund-raising for new programmes. Decision-makers are encouraged to give more attention to developing countries and food security concerns, and to make available more financial support for strengthening research about and in developing countries.

In the early 1980s, agricultural research in Germany had reached an advanced stage of development in terms of capacities, fields and themes of research. Research programmes incurred increasing costs, giving rise to questions of justification for such a research capacity. Among the various reasons for such a downturn in public opinion, the following seem to be of particular importance.

1. Attention has been drawn to the *unbalanced ratio* between contribution of agriculture to the gross national product (GNP) and public expenditures for agricultural research. Large groups of German society do not understand why such a costly research body for agriculture is needed when the sector contributes less than 2% to the GNP and involves not more than 2% of the total working population. Other economic activities contribute at a much higher rate to GNP and offer far more jobs with less public expenditures for research. Decision-makers continue to cope with these crucial questions, and must justify their choice in favour of agricultural research. Sometimes they run short of arguments.

These facts are not new. However, as long as public budgets were not too tight, cutting back budgets was not really a strong concern. For some years, budgetary constraints have been crucial and there is, more than ever, a strong public awareness for the need to restructure expenditures for agriculture, and implicitly for agricultural research. Reforms are targeted at bringing public expenditures in line with the role of the primary sector in the national economy.

2. For a long time, one of the main concerns and targets of agricultural research has been to increase production and yields. Figures on production growth prove that this goal has been achieved. For that reason, people wonder why we need research with such an orientation, because agriculture in Europe and in Germany has already achieved a *high technological level* in the main production areas. Furthermore, *production surpluses* are a burden on the public budget through storage costs, export subsidies, and other measures which are targeted at keeping prices at the required level. If there are unsolved problems here and there that justify research, it is suggested we make use of the private business research sector to find solutions. These same people consider that private business research is in a better position to identify researchable questions, to find solutions in a shorter period of time, and, in doing this, to realize a balance between costs and benefits of research. In some key areas, essential for farmers, private research may be more successful in developing new products and technologies for farmers' needs, specifically in the area of breeding, plant protection, fertilizing procedures and agricultural engineering. These products and services are supplied to a large extent by private enterprise. The question frequently advanced is: for which tasks, beyond the mere control of product composition and of effects of applied fertilizer and plant protection procedures, should public research pay?

3.   In addition to these two critical issues, the research landscape in Germany is facing another problem which has become evident during the past decade: the *demand for university graduates* in agricultural sciences. Demand expressed by public administration, parastatal organizations and private businesses has gone down. Graduates cannot find appropriate jobs in agricultural administration, extension or research, or in farms and agribusiness services that provide farms with inputs or buy farm products. Regional political authorities feel an urgent need to reduce training capacities in order to avoid an increase in trained specialists with poor employment prospects. Within the institutional setting of training and research in Germany, cutting down training capacities entails reduction of research capacities.

4.   The *European integration* process impacts not only on organization and development of product markets, but reaches more and more into other areas. Research is one of the affected areas because research results are tradable goods which are part of the exchange system on the European level. Demand for innovations created in Germany may be satisfied by institutions located elsewhere, even if demand is expressed by the public sector and public decision-makers. With research data so mobile and exchange possibilities and transfer mechanisms improving constantly, research can place its operations almost anywhere. Thus, research institutions should no longer be organized and work only within national boundaries. Other countries may also have an advanced research system with a high specialization in some specific crops or research fields, according to the importance some products have in their own production system. Why not rely on their results instead of maintaining national research capacities for all crops? Such questions are asked by many high-ranking decision-makers.

A second point needs to be mentioned. The European Commission itself is looking for research results. Even if certain services have to be offered in a balanced way between member countries in allocating research funds, they will be forced to take into account cost aspects and to give preference to suppliers with lowest costs. This is a particularly strong constraint for German research institutions because their costs are high compared to other countries within the Community.

5. Last but not least, the *reunification* of the two German Republics marked a turning point in adjusting the agricultural research and university training system. Over a period of 40 years, the former East Germany had gradually set up a large research infrastructure endowed with many scientists. The necessity to incorporate at least some of these capacities into the existing agricultural research system could open the opportunity to adjust the agricultural research system in Germany as a whole. This was at least the declared expectation of the Scientific Council in Germany entrusted with the task of conducting the evaluation and elaborating adjustment proposals.

One may conclude that the overall climate is not conducive to maintaining the inherited structure and capacity for research, and even less for further investments in agricultural research. Decision-makers felt rather an urgent need to compress public expenditures for agriculture in general and for agricultural research in particular. The level and capacities of research as established during three decades cannot be maintained. This is the outcome of constraints and challenges that have occurred since 1990. Decision-makers suggest combining different groups, to strengthen research organizations and to focus programmes on a few key issues of high priority.

New Options and Future Perspectives

Reunification impact on agricultural research

The problem
The political reunification in Germany occurred in a period when some first steps of adjustments in agricultural research and university training were already in preparation or under way. The former East German administration, now under the rule of the Federal Government, had to be endowed with funds and that was not an easy task taking into account the heavy burden that reunification caused for the federal public budget. Under such circumstances, it was expected that agricultural research would be affected by budget cuts, all the more so because the research and training system, including agriculture, was heavily overloaded in terms of staff. The reasons for this included:

1. The former GDR had a large area of arable land and a comparatively high percentage of high-value land, relative to total population.
2. Agricultural production and self-sufficiency were high-ranked targets in the framework of the economic and social development strategy of the former government.

3. The planning process as it was organized needed large inputs of human resources for research, planning, extension and implementation operations. In many cases research functions overlapped with other activities. There was no clear separation between technology development and technology application activities.

4. Research institutes had to contribute to the overall objective of full employment. Even if in the last decade the GDR economy faced shortages in the labour force, university graduates could not easily find an appropriate job. Research institutes were urged to employ these people.

The agricultural research system was subdivided into different types of institutions:

1. The Academy of Agricultural Sciences (AAS) which had under its auspices institutions dealing mainly with upstream-oriented, fundamental research in many fields with a total of 3300 scientists. They were not necessarily involved in university training, but collaborated frequently with universities; some scientists were also teachers at the faculty.

2. Institutes under the direct control of the Ministry of Nutrition, Agriculture and Agribusiness employed 2900 scientists.

3. Universities and affiliated training institutions employed 1350 researchers and teachers.

In 1990 agricultural research and university training employed about 7750 scientists. West Germany, with a twofold higher arable land area, had only 3600 agricultural scientists. Staff reductions were unavoidable. There was a general understanding that university training and research capacities could not be maintained at the existing level, and had to be reduced. The number of institutes and scientists to be maintained in a thematically restructured research system was not determined in advance, as no priorities had been set before the adjustment procedure started. The following objectives and criteria were the guiding principles of the evaluation exercise:

- The existing structure and capacity should be brought down to achieve more or less the West German level in terms of number of scientists per unit of arable land.
- Faculties, research centres and institutes should be given approximately the same shape as institutes in West Germany and elsewhere in the Western world, particularly in terms of physical infrastructure and number of scientists.
- Research units that proved to be of high quality should be incorporated into the new system and consolidated in an appropriate way.
- Research tasks that were funded by public budgets in East Germany, but which were carried out in West Germany by private

enterprises as private business-oriented research, should be privatized.
- Scientists with outstanding records of achievement should be given the opportunity to continue research in their field of excellence.

The decision for such an exercise was made immediately after 1989. It was decided that the structure in place would be funded until no later than 31 December 1991.

Overview on adjustment decisions
The recommendations achieved concerned all types of research and training institutions operating at the beginning of 1990, comprising basic and applied research as well as training. With regard to training, the former East German government had under its auspices some institutions that satisfied the specific needs of a centrally planned economy: researchers, planners and managers in state-owned farms and processing units.

Concerning *universities*, four agricultural faculties (Rostock, Berlin, Halle, Leipzig) and two so-called agricultural high schools (Bernburg, Meißen, in charge of training top executives and managers for the socialist planning and production units) were evaluated. It was recommended to maintain two large faculties, one in Berlin and the other in Halle. In Berlin, the two existing faculties (one in East and the other in West Berlin) were proposed for merging into one large faculty under the heading of Humboldt University located in the former eastern part of Berlin. The former universities in Rostock, Bernburg and Meißen were transformed into higher training institutions ('Fachhochschule'). The Centre for Tropical Agriculture in Leipzig was proposed to be closed without any possibility of absorption by another existing or newly created structure of research or training in the field of tropical agriculture. The total number of staff in agricultural faculties was reduced by about 60–65% (Scientific Council, 1992a).

With regard to *agricultural research outside universities*, the former GDR had about 50 centres and institutions with approximately 6200 scientists (Scientific Council, 1992b). About 53% were working within the Academy of Agricultural Sciences, and 47% in the agricultural ministry's own structure. The breakdown of scientists according to specialization was as follows:

- Soil and plant production   38%
- Livestock and veterinary science   23%
- Food technology (production and product technology)   20%
- Agricultural and rural engineering   10%
- Economics, data collection and documentation   9%

Searching for a new structure for agricultural research, the evaluators considered all options. The results are presented in Table 3.1.

Table 3.1. New structure and capacities of research outside universities.

Affiliation	Number of centres/institutes	Number of scientists
Blue-list institutes	5	270
Federal research under the auspices of the ministry	2 (plus 2 institutes incorporated in an existing centre as a station outside headquarters)	330–350
Regional centres, run by more than one region	4	66–68
Regional centres	5 (1 per region)	250–270
Universities		150

The large number of new blue-list institutes (Federal Ministry of Food, Agriculture and Forestry, 1994) may constitute, to some extent, a new element within the German agricultural research system. In fact, before 1989 not one research institute specializing in agriculture functioned under such a rule. One may conclude that the evaluation procedure marks a turning point in favour of shared responsibilities for agricultural research between several ministries. This may be a chance and a challenge for research institutes, because they will have to respond to demands of different decision-makers with quite different demands for research. On the other hand, a ministry managing a research unit together with other partners must necessarily be ready to specify its own information needs and to determine with the sister institution and the concerned researchers, physical and human resources needed to carry out the tasks.

Second adjustment during the mid-1990s

The reunification process resulted in an expansion of research capacities, expressed by the new land/scientist ratio, as an indicator for measuring and comparing research capacities. Scientists involved in the restructuring process, as well as decision-makers, were aware that any adjustment of that type must be an ongoing process and that capacities and organizations established at the beginning of 1992 had to evolve. The structure was conceived as a provisional, open-structured concept. Everybody was aware that some future adaptations were needed.

Beyond these inherent dynamics of any adjustment, the overall economic development in Germany was such that strong reduction of public budgets became necessary in the mid-1990s. Economists thought that decision-makers underestimated the financial burden and inherent costs of reunification. Public budget reduction also affected the agricultural research system. Starting with federal agricultural research, the Government intended to cut the research budget by 26% by 2005 (Federal Ministry of Food, Agriculture and Forestry, 1996). Implementation could be done in two ways: (i) by asking research centres to produce an adjustment plan including priority setting; or (ii) giving detailed instructions on how and where capacities have to be cut.

Having in mind the modest results of adjustment efforts in the 1980s, the Ministry opted for the second approach, indicating the staff and the number of centres and institutes to be maintained and consolidated within the period indicated above. It will be up to centres and institutes to implement the contraction process through attrition. The target structure of the new and restructured federal research organization is presented in Table 3.2.

Table 3.2. Restructuring plan proposed for federal research centres.

Structure in December 1995			Structure by the year 2005		
Centres	Institute	Staff	Centres	Institute	Staff
FAL (multidisciplinary)	16	850	FAL	9	560
BBA (biological control)	15	682	BBA	9	560
BAZ (breeding)	13	464	BAZ	7	311
BFAV (animal virology)	8	361	BFAV	5	286
BAM (milk research)	6	212			
BAGKF (cereals, potato, fats)	5	174			
BAFF (red meat)	4	103	BFAPE[1]	10	492
BFE (nutrition)	5	180			
BFH (forestry)	8	230	BFH	7	185
BFAFi (fishery)	5	217	BFAFi	4	165
ZADI (information)	1	43	ZADI	1	39
Total	86	3516		52	2598
Scientists		830			703

[1] Research Centre for Food Products.

'Länder' budgets were subjected to similar modifications. Since 1995, yearly university budgets have been, at best, held at the same level in spite of a slight inflation rate of about 2% annually. In some cases, cuts have been unavoidable. In contrast with federal research, administration did not anticipate adjustment measures in detail but left adaptations up to universities and faculties. Berlin is worthy of mention as an exception: the state parliament (Senat), which had to cope with severe financial shortages, announced an austerity programme in 1996 and proposed, in a first consultation round, to dissolve the agricultural faculty which was established 5 years ago as a result of the above-mentioned merging process. In a later stage of decision making, the initial harsh suggestion ended up in a sharp budget cut guaranteeing the survival of the agricultural faculty in Berlin.

Budgetary constraints suffered by regional administration do not have an immediate effect on permanently assigned staff of universities. But it is obvious that even a small reduction of operating funds will affect their abilities (and the motivation) to prepare and to manage research projects. Furthermore, financial constraints may induce ministries to use their medium-term options to influence research priorities and programmes. In coming years, vacancies for professorships may not be filled, at least not immediately. In this way, the whole university research landscape in agriculture may go through a major transformation.

Problems in balancing budgets also affected the Federal Ministry of Education, Science, Research and Technology (BMBF), which disperses research funding on the federal level, with the exception of direct support to agricultural research centres. According to the extent that the Federal Ministry of Research provides funds for research programmes focusing on agriculture (for example through DFG), agricultural research will be affected indirectly. The long-term impact may be crucial as these funds are targeted mainly at promoting young scientists.

Conclusion

The political, economic and social environment of agricultural research in Germany has changed substantially during the past 20 years. Decision-makers who are responsible for agricultural research management, and research institutions and scientists had to cope with problems resulting from reduced funds and modified thematic priorities. These factors will continue to drive future developments and to force further adjustments with respect to institutional development, funding procedures and focal points of research programmes and themes. The transformation of the research landscape is already under way and will continue to be the main concern in coming years.

Taking into account the different factors determining the future of the agricultural research system in Germany, the following hypotheses with respect to future developments may be pertinent:

1. The basic structural characteristics of task-sharing in Germany's agricultural research system between the federal, state and regional

states ('Länder') will remain the same as those which applied in the past. The system will be based on three main pillars: university research, federal research and private research. The different players will continue to make coordination efforts. They will slightly improve the functioning of the system and hopefully develop several jointly managed programmes. Nevertheless, experience with the ATSAF initiative shows that such attempts at overcoming the scattered research system reach their limits quite quickly.

2. Budgetary funds for public agricultural research systems will decrease, at least in real terms, particularly for the federal research system. Furthermore, the allocation procedure may be gradually modified so that special and project-related funding takes greater importance at the expense of global institution funding. Users of research results will benefit from such procedures and research groups and individual scientists will get incentives for more user-oriented research.

3. The thematical orientation of research will be guided by new challenges. Four topics will be of particular interest in coming years:

- Environment-related research and resource conservation, which will affect research in the fields of plant and animal breeding, plant physiology and pathology, rural engineering and economic and social research.
- Biotechnology and particularly genetic engineering (Council for Research, Technology and Innovation, 1997).
- Transformation problems and restructuring procedures of the agricultural sector in Eastern Europe as an interdisciplinary task with topics that are derived from concrete decision-making problems.
- Impacts to be expected on German agriculture and agricultural marketing from globalization and possibilities of further adjustments.

The agricultural research system in Germany is going to shift towards a system with a stronger focus on some research programmes of high priority, with demand-driven and user-oriented topics, and with a higher flexibility of human resources management. The guiding principle with respect to research funding for all research institutions under the public auspices will be: less regularly provided funds and more competitive bidding for project-specific funds offered by a wide range of public (ministries) and private institutions. Ministries have been urging research institutions to make every effort to reduce costs and to seek innovative ways to raise their own funds.

Research Investment Strategies in the Americas: Lessons Learned

II

The United States

4

Neville P. Clarke

The Agriculture Program, The Texas A&M University System, Centeq Research Plaza, Suite 241, College Station, Texas 77843-2129, USA

Introduction

This chapter reviews the US experience in funding and managing agricultural research, where the focus has been on the relationship between the state agricultural experiment stations and their federal sources of funding. The premise is that there are substantial commonalities in the hierarchical structure of this system with that of donor–developing country relationships, and that some of the earlier US experiences are now emerging in international agricultural research.

In presenting the US experience, there is the risk of seeming to presume that the USA is better than others or that we can teach and others can learn. In fact, we wish to present the US experience with considerable humility, believing that others may benefit from our mistakes, as well as from the things that seemed to have had some utility in our environment. Many of our observations restate well-recognized management principles, couched in the vernacular of our experiences. We do not suggest how others should operate, but rather share our experiences with the hope that, with adaptation, some of them may have utility.

The US Land Grant University System

Land grant universities have undergone substantial change over the past 20 years, moving from a support base that was relatively unconstrained and ongoing to one that has been downsized and is increasingly dominated by more focused and restricted funding. The agenda has moved from one dominated by production goals to one that involves more substantial environmental and postharvest components. New opportunities in science have resulted in a substantial shift in the research portfolio towards basic science. The changes faced by the land grant universities are, in many respects, similar to the changing agenda that appears to be facing the International Agricultural Research Centres (IARCs), national agricultural research systems (NARS), and others involved in international agricultural research and natural resources development. This section summarizes these changes as background to the more detailed assessment of their impacts, and the ways in which the land grant universities have gone about dealing with change.

The network and its regional and national linkages

Federal legislation created land grant universities (LGUs) in each state. Later, state agricultural experiment stations (SAESs) and cooperative extension services were incorporated as parts of these universities. All formerly black universities were incorporated into the broader system, as were colleges for Native Americans more recently. A key element in establishing and maintaining effective networking between the state institutions has been 'formula funds', a relatively small amount of money appropriated by the federal government for each state institution on a formula related to the size of the state's agriculture. Management of these funds has provided the 'glue' that holds the system together and facilitates forward planning of the interface between federal and state systems.

The interests of the states are consolidated at regional and national levels for planning, cooperation and advocacy. State institutions, with primary motivations to serve individual states, continue to find strong common interests which help define the regional and national agenda, and form the basis for a continuing loose network of cooperating states along with the federal institutions which together form the national system.

Comparative and competitive advantage and synergies

The SAESs (Clarke, 1984, 1993a; Miller, 1993) in those states where agriculture is a major part of total economic activity are funded well by their individual legislatures; the combination of federal formula and state funds provides a continuing stream of funding for research similar to that which existed in the CGIAR when unrestricted core funds were the dominant resource. When these resources were less constrained and larger, they provided long-term stability, the capability to attract good scientists, and the ability to attract and lever external funds. SAESs tend to mirror the overall growth in excellence of their parent institutions, so that a stratification has occurred over time in this respect among states. Although substantial changes have occurred, this system continues to provide the capability to serve the needs of an expanding clientele with multiple agendas.

Changing US perspectives between domestic and international agricultural research

There has been a substantial change in attitude of the LGUs and their stakeholders about the role of the SAESs in international agricultural research over the past 10 years (McMillan, 1997; Moser, 1997). It is now generally recognized that US agriculture is part of international agriculture and must compete in the global marketplace. Further, agribusiness and farm organizations acknowledge that international agricultural research, by contributing to economic growth in developing countries, creates or expands markets for US products. US policy-makers increasingly recognize that food insecurity is a major factor in global unrest and that US investments in international agricultural research help to reduce such insecurity, creating 'preventive diplomacy' as a less expensive and more humanitarian alternative to crisis mitigation.

There is close correspondence of the stratified elements between the US network and the Global Agricultural Research System (GARS):

US network	Global system
Federal government	CGIAR
State governments (SAESs)	National governments
Regional research programme	Regional research consortia of developing countries
Federal Agricultural Research Service	International agricultural research centres
Grants and contracts to individual SAESs	Bilateral projects

Federal funding to the SAESs is targeted towards objectives that are national or at least regional in scope. State legislatures fund research principally targeted to benefit the industry in that state. Some federal funds that come to individual SAESs can only be spent on regional cooperative research. Individual federal agencies fund individual SAESs to perform specified research to address their specific needs. In practice, there is a substantial overlap in the performance and outcomes of research sponsored by these various mechanisms. However, the stratification provides a focus and direction which is generally beneficial.

The global system has interesting similarities to the US system. The CGIAR, with its global perspective, corresponds to the federal government. Overall outcomes are intended to be applicable to more than one country; the general mandate for the IARCs (which correspond to the federal research facilities in the USA) is to conduct more upstream research, leaving specific applications to individual NARS. The US regional research programme, which has been quite successful in achieving its goal of promoting more efficient use of resources through cooperation, is somewhat analogous to regional and ecoregional organizations emerging in international agriculture. It seems to offer some useful experiences as investment strategies for regional cooperation are defined among developing countries (Clarke, 1996b). The regional cooperative research model is considered in detail later.

General overview of conceptual and organizational correspondences for the global system

Although site specificity is a pervasive characteristic of agricultural production, the SAESs have found, especially with recent resources constraints, that there are major efficiencies to be gained through careful planning and execution of cooperative efforts. Customers and directors are increasingly recognizing that not every experiment has to be replicated in every region of every state. Important research is underway to improve the precision and accuracy of extrapolating results from one location to another.

Resource constraints have motivated individual directors of SAESs to create innovative linkages with their counterparts in adjacent or even non-contiguous states. Regional consortia have been formed to address areas of common interest either along commodity or resource lines of study. The key to implementation is careful planning and commitment of resources to a common agenda, with policy-makers involved in the commitment. Special care is required to minimize transaction costs. Regional cooperative efforts work best when they involve commitment of individual scientists from the outset, but they can be stimulated by creating funding streams that can only be used for cooperation. Private sector cooperation and enhanced linkage with extension systems have also proved to be rewarding.

CGIAR has a major advantage in its ability to consult and agree on a general agenda, in contrast with the continuing relative inefficiency resulting from multiple US federal agencies separately funding highly related research in universities, intramural programmes and industry with little or no effective coordination (Petit, 1996).

Priority and Policy Setting at State, Regional and National Levels

The primary mission of land grant universities and their related agricultural research and extension functions is to meet the needs of individual states. With ongoing institutional support from the federal government, there is also a mission to address national needs. The agriculture and natural resources programmes in land grant universities are linked into a quasi-independent network of related activities which share a common agenda at the national level. This is not unlike the relationship between national and subregional/regional programmes in the developing world. The methods that have evolved in the USA for seeking common agenda and agreeing on shared priorities are summarized in this section in the context of the emerging global system.

Seeking consensus among stakeholders

Developing and maintaining meaningful consensus among stakeholders is tedious, ongoing and necessary. The closer to the farmer, the more specific and explicit must be the consensus about priorities and programmes. Given that the individual SAESs are primarily motivated to serve the needs of individual states, it is remarkable that there is a good consensus among states about the most important things to be done in research among both the directors of SAESs (Clarke, 1996a) and their various constituencies at regional and national levels (Anon., 1995; Clarke, 1996b).

Most recently, LGUs have published a consensus-based strategy on agriculture and natural resources which defines a broad agenda for research, extension and higher education (Anon., 1987, 1994; Clarke, 1996a). Almost 400 leaders from agricultural industries advised LGU administrators on this agenda. In taking the broad agenda to specific action plans, the key elements in achieving consensus include:

- Resource limitations as a motivation to cooperate and reduce redundancy.
- Inability to achieve all goals, which forces priority setting.
- Recognizing institutional comparative advantages as a matter of necessity.
- Overcoming state and institutional 'pride'.
- Getting the right people together at the planning table.
- Continuing the dialogue until convergence on meaningful plans emerges.

Setting outcome-driven goals and objectives

The US agricultural research community continues to learn how best to engage in a planning process that focuses on *outcomes* rather than *outputs*. Increasing customer participation in agenda setting has helped this focus, but often places too much emphasis on short-term adaptive studies. The Government Performance and Results Act (GPRA) mandates that all parts of the federal government set goals against which progress can be measured, and that progress be assessed on an annual basis. The intent is to direct resources to agencies and goals where progress is demonstrated. In practice, it is possible to set broad goals at the national level, but specific goals and measures of progress depend on more specific goals at the state level. The USDA has initiated the process of meeting the requirements of GPRA, but has not yet demonstrated how well it will work. As is the case with developing countries, donors and the CGIAR in general, there is the unremitting question of how much of the total resource should be allocated to assessment.

It is planned that aggregates of state level results towards a common national goal will be used to assess national progress. There appears to be a substantial correspondence with this and the goal of the CGIAR to assess and evaluate the impact of its research, where common system level goals and objectives (TAC Priorities and Strategies) are configured so that progress across centres can be assessed.

Targeting fundamental research

The science establishment in the USA has reluctantly moved towards a more explicit targeting of fundamental research as opposed to open-ended scientific endeavour. Agriculture has moved further in this respect than other parts of the establishment. A continuous balancing act is in play where the attempt is to ensure a clear definition of target outcomes without unduly constraining creativity in the pathway towards the outcome. If one considers the overall state–federal investment in the SAESs, considerably less than 10% of the total is invested in fundamental research. It is estimated that the optimum would be for not less than 25% to fall into this category. Fiscal pressures have led state legislatures to focus on specific goals for new funds, often with short-term goals.

The USDA initiated the National Research Initiative in 1989 to provide new funds for targeted fundamental research (Clarke, 1984, 1993b; Hullar, 1989). It is funded at about US$100 million year^{-1}, having fallen short of the goal of US$500 million year^{-1}.

In the case of international agricultural research, there is a distinction between fundamental research done in advanced research institutes (ARIs) that contributes to capabilities in developing countries vs. fundamental research done in NARS for their own purposes. The latter tends to be directed more towards focused goals. Maintaining an explicit and obvious linkage between fundamental, applied and adaptive research provides the best assurance of continuing utility and relevance.

What's needed vs. what's possible

Whether it is the US agricultural research agenda or those of other donors or developing countries, often the highest priority problems are not perceived to be amenable to research solutions. The USA continues to struggle with this dilemma in the decision-making process. Better planning based on outcomes is helping to focus on what is possible versus what is needed. It's often not a question of black and white, but shades of grey.

Dealing with the politically possible agenda

The process of broad consensus seeking, with multiple inputs into the process, frequently forces research managers at all levels to deal with the 'art of the possible'. Politically driven decisions, explicit goals of individual investors/donors and compromises within universities to cover teaching-based disciplinary objectives all tend to dilute the objective decision-making process that most managers would like to employ. Our experience suggests that an effective and persuasive outcome-oriented strategy can frequently be used in such debates to identify and (sometimes) play down proposals that are too narrowly focused, and to show the trade-off costs of taking up one agenda at the expense of another.

An operational perspective on utility of economic models

Using well-constructed economic models as one input to the decision process is proving to be highly advantageous, in terms of *ex ante* analysis of proposed research and *ex post* assessment of results. Such results are of special importance in the advocacy of research with policy-makers and decision-makers. The modern-day research administrator functions in many respects as a facilitator or communicator between various levels of decision-making to ensure that overall strategies are turned into actions that have the best probability of achieving desired outcomes.

The state of the art in *ex ante* and *ex post* estimation of the impact of research is rapidly evolving. For example, the new impact assessment programme at Texas A&M is integrating several related ongoing tools to create a new capability. The network of related models uses a GIS framework with a variety of natural resource, census and geopolitical databases to facilitate analysis and display of results from network models that describe the biophysical function of various plant and animal cropping systems; sectoral models that describe the impact of research on yields, markets and regional distribution; environmental models that predict outcomes of the application of new technology on natural resources; input/output and demographic models that predict the consequences of new technology on communities and families; models that predict the consequences of various inputs, including new technology, on probability of survival of the representative enterprises; and a final GIS-based synthesis that presents the results of these analyses on specific locations.

Adapting to Change in Major National Issues and Opportunities

Land grant universities have undergone a major change in their agricultural research and extension agenda, which results from changing customer needs and demands and new opportunities in science to meet continuing and newly recognized mandates. A description of these changes and the response of the land grant universities will offer insights into the management of evolving agenda for international agricultural research, especially with respect to the CGIAR and its donor members.

Biotechnology and business

In most larger LGUs, there has been substantial investment in biotechnology in the 1980s and 1990s; often this has involved a re-investment of existing resources, sometimes new funds have been appropriated by states or acquired through industrial or competitive federal grants. While mostly positive, this shift has created some shortfalls in disciplines that remain critical to the total portfolio. For instance, plant and animal geneticists working at the molecular level have increased in number, whereas traditional plant and animal breeders are becoming difficult to find and recruit.

The SAESs have learned some hard lessons in the past 15 years about biotechnology, intellectual property, industrial partnering and resources. The major messages are:

- The opportunities in biotechnology are real and rapidly emerging in applications today.
- The developing world stands to benefit much the same as the industrial world.
- Delivering products of biotechnology usually requires an industrial entity for terminal development, regulatory clearance, production and marketing; without this, good technology will sit on the shelf unused.
- For industry to be involved, they must be able to make a profit.
- It is not inherently wrong for industry to make a profit.
- To justify their investment, industry must be able to protect intellectual property.
- Those using biotechnology methods to develop technology for developing countries must consider early protection of intellectual property or it will not be picked up by industry.
- There are contractual methods that can be used to ensure 'due diligence' and to minimize costs to developing countries.
- The USA has gone from a situation where there was a stigma associated with industrial partnering to one where this is expected and encouraged by our stakeholders. In regard to contracts/partnering with the private sector, general templates do not usually work; you have to negotiate each deal individually. Public institutions must sharpen their negotiating skills to protect themselves in this arena.

Concern for natural resources and environment

Public concerns about the natural resources/environmental agenda have increased steadily in the USA and are linked to growing concerns about diet, health and general well-being. These concerns are strongly reflected in public polls, resulting in modification of political directions proposed by Congress in recent years. To predict the trends in the USA, we frequently look to countries in central Europe, where these concerns have resulted in substantially greater changes in policies and practice. Increasingly, SAESs must be concerned about natural resource and environmental consequences of all R&D. Methods to factor environmental consequences into the overall process of agenda and priority setting for research are still emerging.

The sustainable agriculture interests in the USA continue to grow and are having impact on the research agenda in the SAESs. Major manifestations are: (i) the increased emphasis on production systems with the development of methods and products that lend themselves well to CGIAR ecoregional initiatives in sustainable crop–livestock systems; and (ii) methods and practices to minimize the purchased inputs for production. Even though there is criticism of the damage resulting from intensive production practices in some developing countries, in many other locations present use is so low and needs for food so great that increased use of chemicals can and should be encouraged.

The imposition of the environmental agenda of the North on to the countries of the South, where the concerns for food security and poverty alleviation are greater, is a sensitive issue. However, our experience suggests that early and continuing sensitivity to these issues in research agenda setting can result in decisions that contribute to environmental protection and have minimal or even complementary impact on goals related to production.

Criteria for Making Adjustments to Changing Resources

Creating new agenda, especially in an environment of shrinking resources, is challenging at best. The methods that have been used to manage change have varied considerably between institutions. Those institutions that have been proactive and responsive have survived and done relatively well; those that did not are losing ground. There are valuable lessons to be learned from our mistakes and more appropriate responses to external forces.

Re-evaluating priorities in a multi-dimensional stakeholder environment

Lessons to be learned from the US experience are the importance of:

- broad consensus seeking;
- maximum transparency in the process being used;
- having the vision and courage to escape the current paradigm where real change is needed;
- developing and using as objective a method as possible;
- maintaining the posture of letting seasoned experience and creativity of scientists prevail at the end of the day.

Downsizing and right-sizing: goals, methods, constraints

In the USA, two models have emerged to bracket the continuum of management solutions with downsizing; the overall science capacity of the SAESs has been reduced over the last 10 years by about 25%: (i) employment has been preserved for as many staff as possible, with ratios of salary to total expenses sometimes exceeding 90%; and (ii) preservation of an appropriate salary to total budget ratio of approximately 75%. Unless downward shifts are definitely determined to be transient, the former method involves a slippery slope to oblivion. The harder but more sustainable approach is to prioritize and downsize while maintaining the capability for those remaining scientists and programmes to maintain a critical mass of resources for their work. Mandatory downsizing is no fun, but it is important to remember that adversity creates opportunity. Decisions that are programmatically desirable but otherwise *unthinkable* with ample resources become *possible* when resources are sufficiently constrained.

Traditional vs. new missions and clientele

As the NARS and IARCs, as well as the ARIs, continue to encounter more constrained and restricted funding, the challenge of maintaining mission focus becomes more daunting. In the USA, the agenda for the SAESs expanded as resources were reduced, a situation not unlike that encountered in international agricultural research. New clienteles bring interests in parts of the pre- and post-food production system, more explicit concerns about natural resource use and management, and mechanisms to more effectively accomplish technology transfer to farmers. Experience suggests that as the global system takes shape, the opportunity to co-mingle funds from multiple sources offers a mechanism to sustain momentum of key core programmes in the transformation from 'hard to soft' money.

Experience also suggests that research managers and policy-/decision-makers in developing countries must be prepared to be considerably more flexible and agile in reacting to their surrounding environments; the pathway to the strategic goal may not be as direct as desired, but general directions must be maintained as options are exercised. There is a need to minimize bureaucracy and transaction costs and to be creative in developing new investment and operating paradigms.

Stratified Strategic Planning: Implications for the Global System

Strategic planning has been a central part of communication and setting the investment strategy at state, regional and national levels in the US land grant system over the past 15 years. Most recently, an integrated plan for action on agriculture and natural resources in the land grant universities has been developed (Clarke, 1996a) and is being used as a vehicle for multi-layered communication throughout government and in LGUs. As noted earlier, there are correspondences between the US network and the emerging global system which may make US experiences useful as background (Anon., 1996).

Useful planning must be a careful mixture of bottom-up and top-down approaches. The ideal situation is when these two processes are iterative. Problems and the 'art-of-the-possible' in gaining support from external stakeholders often mean that the major issues are defined by the external environment. Opportunities more often arise internally from the creativity of scientists who are also aware of the needs of the customers they serve. In the USA there is an increasing drive for clientele to become involved not only in agenda setting, but also in experimental design, which often takes them well past their level of meaningful insight. It is important to ensure that the first level of detailed planning occurs at the lowest organizational level. State plans form the basis for regional plans and regional plans form the basis for national plans. We suggest that similar logic would apply to the strategy-setting process for the global system.

It has begun to be accepted that research and the SAESs cannot operate in a vacuum, but must increasingly depend on counterparts in the cooperative extension system and industry to ensure that the products of research are put to use in a timely manner. Joint strategic planning is helping us to bridge the gaps and identify areas of most productive cooperation and interface. In the USA many strategic plans end up sitting on shelves or in filing drawers, serving only the purpose of documenting communications that occurred between drafters. As we have undertaken to make such plans more outcome oriented and our customers have become more involved in priority setting, plans have begun to be vehicles for advocacy of resources and frameworks against which progress can be measured and results communicated to users.

Financing Agricultural Research: Multiple Sources/ Multiple Objectives

Changes in the mechanisms for funding of agricultural research and extension in the US system, and the resulting programmatic changes that occurred in the 1980s, are being replayed now for the IARCs, NARS and other players in international agricultural research and development. Valuable lessons can be learned from the appropriate actions and mistakes that have been made in the USA in the allocation

of shrinking resources to new agenda. The concern over the dominance of short-term goals to strategic research and the loss of institutional stability needed to sustain comparative advantage are cases
in point.

Maintaining mission focus

In the US experience, when the equivalent of restricted project
funding exceeds about 35% of total budget, the ability to retain focus
on the mission goals of the institution and the commitment of scientists to these goals is compromised. In the case of one large SAES, the
restricted funding was less than 10% in the mid-1970s; it is now
greater than 50%. At the same time total funds have more than doubled
in real dollars. That institution is doing more research now than it was
then, but the ability to focus the research on mainstream problems of
the state has been substantially compromised. The equivalent of unrestricted core funding has been used to support the infrastructure
which enables restricted project research on topics that are frequently
not mainstream to the goals of the institution. Restricted funding often
covers fundamental research that may be only remotely related to
mission, or it may cover applied research that might better be done by
industry or the extension services.

A similar situation is emerging in the IARCs where unrestricted
core funding has been reduced substantially. The loss in flexibility, the
tendency to compromise longer-term upstream research and the ability
to maintain institutional focus appear to be following the same pattern
as occurred in the SAES example 10 years ago.

Restricted funding to support core programmes

Earlier US experience would suggest that the real challenge to management in this situation is to find restricted funding that can substitute
for unrestricted funding. For customers (donors) the challenge is to
develop ways to persuade policy-/decision-makers of the merits of
investing in a balanced portfolio that supports the entire spectrum of
research application. Communication and meaningful planning are
common mandates for both performers and supporters of international
agricultural research. Incentives to encourage better balance might be
sought through distribution of World Bank funds on other than a
formula basis driven by centre income from other sources.

Preserving comparative advantage as a first priority

In the USA the private sector has increasing capability eminently responsive to customers with highly specific research needs and tight schedules. As other kinds of resources have been reduced or vanished, universities and other publicly supported entities find themselves competing for these short-term funds. Another slippery slope to oblivion is when such public institutions lose their institutional comparative advantage to perform well in the kind and quality of science in which they have excelled and find themselves increasingly unable to compete in the more commercial environment. There is a distinct danger of this occurring in the CGIAR community; in fact evidence can already be found that this is occurring.

The message is that, even in a downsizing resource environment, it is critically important to define and sustain the areas of institutional comparative advantage that will keep the institution in business. Recognizing that, in this environment, research institutions cannot be good at everything relevant to their mission, hard choices need to be made and outsourcing and partnering used to fill in the gaps.

The Regional Cooperative Research Model

As the global system looks towards methods to promote and sustain cooperative activities, it is suggested that the US model for regional cooperative research among the SAESs may be quite useful. Seed money from the federal government, which can only be spent on cooperation between states, is coupled with matching funds (or more) from other sources to address problems of shared regional importance between states. The system has worked well, albeit with some problems which might be avoided by others.

The procedure

An amount of money equal to about 25% of the formula funds appropriated to the SAESs by USDA is mandated to be spent on co-operative regional research. The decision as to how to invest these funds rests with the SAES director. Regional associations of experiment station directors plan and approve regional projects (with oversight from USDA) and, in turn, each director determines the amount of his share of regional cooperative research funds that will be invested in each approved regional project. Directors may spend these monies on projects in other regions of the country. Such projects have regional administrative advisers and mechanisms for reporting and evaluating results.

Decisions at the local level

The two key management points in this strategy are: (i) some monies must be spent cooperatively; and (ii) the decision on distribution of these funds rests with the same director who has overall responsibility for the total formula funds from USDA and for the complementary funds that come from state appropriations or other sources. The result is that the mandated regional cooperative funds are increased by a factor of 5:1 to 6:1 with funds from other sources in the four regions of the USA. The most consistent problem with this programme is that regional cooperative funds tend to become embedded in the ongoing programmes of the individual SAESs and in salaries of faculty members presumably doing research on these projects. In some states, the ability to truly mobilize these resources to meet new challenges and take advantage of new opportunities is limited.

US experience would suggest that a funding paradigm such as this would be useful to promote cooperative research between NARS in regional efforts such as The Association for Strengthening Agricultural Research in Eastern and Central Africa (ASARECA). It might also be a useful mechanism to promote collaborative research between IARCs on systemwide and ecoregional programmes.

Centres of excellence

As evolution of this system continues in the USA, states are moving past cooperation and considering consolidation of regional research at one location. Overcoming political and other stakeholder considerations (not all states have all capabilities and state funds must cross state lines to contribute to the common good) is proving to be challenging. This is a situation that corresponds closely with what is occurring in the global system and the IARCs. Establishing 'centres of excellence' which are mutually supported and used by partners with common needs is a development that seems to be emerging.

Criticality of good planning

Our experience in the USA suggests that one of the critical elements in making such things happen is to enhance the quality of planning and communication between concerned parties. New methods using spatially referenced analysis to define geographically common areas, coupled with economic models that allow *ex ante* assessment of local, national and regional impact will contribute significantly to the precise definition of goals and objectives for such international partnering.

Building Interdisciplinary Research Teams

As NARS and others in the global system move towards multi-disciplinary research using the systems approach, there will be utility in understanding how the US land grant system is dealing with the challenges of moving from disciplinary to transdisciplinary research, and to consider the various methods that are being used to promote and sustain this approach. Such methods are particularly relevant to the research of IARC–NARS–ARI teams working at the ecoregional and global levels.

Promoting collaborative research

Multidisciplinary or transdisciplinary research is inherently more complex than 'single author – single discipline' studies. Accepting the need for both, there is increasing recognition that creating outcomes usable by farmers requires the former. Creating motivation and commitment to transdisciplinary research has been challenging in the US system. Creating resources committed to such research is an excellent mechanism to promote thinking and mutual commitment between science partners. In our experience, it may often be sufficient to provide start-up funding for such efforts which, if successful, are sufficiently rewarding both intellectually and financially to be sustained thereafter.

IARCs have an inherent advantage

The CGIAR centres are particularly well placed to promote transdisciplinary research, since their administrative structure is usually not set up along strict disciplinary lines. National research institutes in developing countries also have more flexibility in organizing interdisciplinary research than do universities. Linking multiple centres and multiple NARS together into ecoregional or systemwide programmes offers complex challenges where all these difficulties can be exacerbated unless particular attention is paid to minimizing transaction costs.

Rewards

To sustain interdisciplinary research, collaborators must perceive both personal and institutional rewards. In the USA we are still working towards creating a peer and administrative reward process that recognizes interdisciplinary research as being 'scholarly' and of sufficient quality to merit reward. In the USA there has been a significant increase in 'centres and institutes' within university communities to provide a vehicle to facilitate interdisciplinary research with minimal transaction costs. Many of these are 'centres without walls', virtual institutions that facilitate cooperation and provide visibility to the collaborative effort with donors and customers.

Effective Advocacy

The need to be more effective advocates for support of agricultural research and development is shared by industrial and developing countries. There have been some successes over time in the US system which would seem to have relevance, at least in principle, to what is needed by the international development community in general. The following general principles are perhaps noteworthy:

- Identifying stakeholders and communicating with them.
- Developing and maintaining the appropriate image.
- Linking technical and resource plans.
- Building in technology transfer at the outset.
- Methods to visualize and accept accomplishments.
- Linking goals to outcomes.
- Keeping the faith.
- Protecting fundamental research with related terminal outcomes.
- Communicate, communicate, communicate.

Summary and Conclusion

- A similar hierarchy of organizations and funding sources exists between the US and international agricultural research systems.
- Multiple sources of funding, including state and federal, offer opportunities to balance the investment portfolio and, through multiplicity, to ensure stability of ongoing core programmes.
- There is growing recognition among US stakeholders of the importance and pay-off of US investment in international agricultural

research with emphasis on benefits of mutually beneficial partnering of US institutions with IARCs and NARS.

- Resource constraints encourage new partnering, including establishing centres of excellence which serve multiple political constituencies.
- Multinational and regional research requires developing consensus on common goals and priorities.
- Agreements to establish centres of excellence between existing political entities require perception of mutual benefit, inability to provide all things to all people, and recognizing and using institutional comparative advantage.
- The focus of strategic planning among institutions must be on outcomes rather than outputs.
- Strategic planning should be an iterative process between top-down and bottom-up methodologies. Institutional planning should be a bottom-up process.
- Fundamental research must be somewhat more targeted to survive in constrained resource environments.
- Real world strategic planning of publicly sponsored research requires dealing with the compromise and the 'art of the possible'.
- Models predicting *ex ante* and *ex post* impact of research are helpful as planning tools, but not as devices to make decisions.
- Biotechnology has arrived; the interface of public and private sector requires special attention but is necessary and positive.
- Natural resources and environment are now part of the total production paradigm and this must be reflected in research; the environmental impact of production research must be considered in priority setting and planning.
- Re-evaluating priorities in changing environments and downsizing where necessary are facilitated by thorough communication and a sound outcome-oriented strategy.
- Adversity creates opportunity in programme redirection resulting from reduced resources.
- Constrained and restricted funding makes maintaining institutional mission focus more difficult; the tendency is to be forced towards shorter-term splintered research with donor funding driving the agenda.
- Preserving institutional comparative advantage is of primary importance in establishing a strategy for a constrained resource environment, assuming that the comparative advantage being preserved is relevant to the mission.
- The US regional cooperative research model would appear to offer substantial advantage as a model for creating multilayered funding and implementation of regional research in and between developing countries.

- Building and sustaining interdisciplinary and transdisciplinary teams for problem-solving research, especially systems research, is necessary; methods for accomplishing this are difficult.
- In the present resource environment, effective advocacy of programmes is pivotal. Good planning, ongoing communication and satisfied customers are key ingredients.

Latin America and the Caribbean

5

Barry Nestel[1] and Matthew McMahon[2]

[1]*Little Goldwell Oast, Goldwell Lane, Great Chart, Ashford, Kent TN23 3BY, UK;* [2]*LAC Technical Department, World Bank, 1818 H Street NW, Washington, DC 20433, USA*

Era of Food Self-Sufficiency: 1950s to 1975

Commodity-specific and broad-based agriculture research stations were first set up in the Latin American and Caribbean (LAC) region over a century ago. Research was generally conducted on a limited scale prior to the late 1950s when the Hispanic countries, with support from both the US government and private foundations, began to address the development of science-based agricultural growth and to consolidate their research stations into National Institutes for Agricultural Research (INIAs). The first of these to be established was INTA in Argentina in 1953, and by the time EMBRAPA was created in Brazil in 1973, nearly every country had its own INIA. Although each of them possessed unique characteristics, and care has to be expressed regarding generalization, there were, especially in the early years, a number of characteristics that were common to many INIAs.

All of them tended to follow the same model in being affiliated with the Ministry of Agriculture and in having an import substitution strategy that focused on applied and adaptive research on basic food crops. A similar approach was followed by the smaller non-Hispanic former colonial territories of the Caribbean. The emphasis on food self-sufficiency, later coupled with social equity, continued throughout the 1960s, although during this period some INIAs achieved a measure of autonomy within the Ministry of Agriculture. The INIAs, however, tended to remain Ministry-dependent, in that they were subjected to bureaucratic procedures and low public sector salaries, both of which limited their autonomy. By the time EMBRAPA was created these problems had been recognized and it was given not only more autonomy and flexibility but a mandate to function as the heart of a National Agricultural Research System (NARS), rather than as a stand-alone INIA.

During the 1960s and early 1970s many INIAs experienced an explosive growth in staffing and funding. Much of this was linked to the heightened interest in agricultural research of certain donors, particularly the Inter-American Development Bank (BID) and the World Bank (WB). Together with the United States Agency for International Development (USAID), the two banks have invested over US$5 billion in agricultural research and research-related projects in the region since the late 1950s. These three donors have also provided a major part of the funding for the three international agricultural research centres (IARCs) located in the region (CIAT, CIMMYT and CIP).

The strategy followed by the INIAs assumed that agriculture could be modernized by transferring and adapting technology from industrial countries and by integrating farmers into the market economy. Agricultural research and technology transfer were seen as a public good. Institutional organization and market development for agricultural inputs and products were starting to be developed in Latin America, and the state's role was widely recognized as being to promote social and economic development (Trigo, 1995).

One effect of the above was that the INIAs tended to focus on short-term production problems; thus, they generated little basic or strategic knowledge on how to keep national agriculture competitive internationally. INIAs seldom had effective links with the private sector and, in some countries, government policies actually discouraged the development of private sector research. At the same time, universities were conceived of primarily as training institutions, usually under the jurisdiction of the Ministry of Education. Since the early part of the century many of them had been autonomous, and although about 80% of the total funding of both public and private universities was derived from government grants, they were seldom accountable to governments for their expenditures (Winkler, 1990). Few universities sought external research grants as this was, at that time, seen as a loss of independence. By the 1970s, however, universities in Chile and Brazil had broken away from this mould. In Brazil some even went so far as to link promotion to research performance. Despite this, few LAC universities made a significant contribution to agricultural research in the region at this time.

Decline of the INIAs: 1975–1990

By the late 1970s it became apparent that many INIAs had grown beyond the stage where they were sustainable without continuous donor support (Antholt, 1994). This situation was exacerbated by the fact that the INIAs failed to adapt their traditional input supply strategy to the changing political, economic and institutional circumstances of the 1980s and 1990s (McMahon, 1992). As a consequence of this, many INIAs have faced recurrent institutional crises during the past 20 years and, apart from Brazil and Uruguay, this situation is still prevalent (Pritchard, 1990; Trigo, 1995; Echeverría *et al.,* 1996a,b).

These crises were, and still are, usually associated with a shortage of funds due to both a reduction of donor support and a failure by the public sector to either fill the gap or, in some cases, to even maintain past levels of support. The INIAs failed to use their new resources to modernize their structure and organization in areas such as policy formulation, programme planning, priority setting, human resource planning and broadening the base for their financial support. Little effort was made to develop linkages with policy-makers and limited success was achieved with technology transfer. Given this situation, all the prime donors in the region have expressed concern about the sustainability of their investments in agricultural research (Pritchard, 1990; Byrnes and Corning, 1995; Inter-American Development Bank, 1992; Alex, 1996; Echeverría *et al.,* 1996a,b) and although they have continued to provide assistance for projects focused on institutional change, they have reduced their level of support for research on production with some funding being diverted to research on the conservation and management of natural resources.

National governments have been concerned about the management capability of the INIAs; their failure to deliver sufficient new technology to small and medium-sized farmers; and their poor record in developing appropriate strategic and operational processes and supporting institutional linkages to meet the needs of a changing and demand-driven agriculture sector. Despite these concerns the INIAs have continued their traditional supply-driven programmes focused on self-sufficiency, often using a limited human resource base to work on a large number of crops without having a critical mass of researchers devoted to any single crop (Lindarte, 1995). Their impact has been further weakened by the absence of priority setting in the research programme and the lack of farm level evaluation of results. Additionally, few INIAs have developed good links with policy-makers, who often appeared not to understand both the public good aspect of much agricultural research and its long-term nature. Thus, research budgets have not only been unstable, but have also declined, sometimes suffering year to year variations that have led to significant staff and programme losses (Byerlee and Anderson, 1995).

By 1993 the agricultural technology system within LAC included approximately 100 public and private institutions, staffed by over 10,000 professionals, with an annual budget of about US$900 million.

The rise in research staff numbers, accompanied by static or declining annual budgets, resulted in a fall in the total expenditure per researcher of 40% during the 1980s. In some INIAs up to 90% of the budget was used for personnel costs, which severely curtailed the amount of research that could be carried out, limited the potential impact of the INIAs and further diminished their credibility. In situations where only 10–15% of the total budget was available for operating costs a reduction of only 5% in the total budget has had disastrous consequences for research activities.

Under the pressure of national structural adjustment programmes, a number of INIAs attempted to accommodate budget declines by downsizing their staff numbers, particularly with respect to support staff. This took place, for example, in Bolivia, Colombia, Jamaica, Peru and Ecuador. Since 1991, budgets have been systematically reduced when adjusted for inflation in Argentina, Brazil and Mexico. In Peru a number of experimental stations have been transferred to other institutes.

The US Land Grant University model featured in the early organizational structure of many INIAs, but a common feature of downsizing has been to remove technology transfer from their mandates. Conflicts and competition between research and extension, coupled with poor results from extension, led many governments to establish separate organizations for extension. Few have achieved the expected results, and many countries have moved towards a model in which the INIA is responsible for research and for transferring its findings to intermediary agencies such as NGOs, development corporations, producer organizations and rural development projects, which all serve as independent extension agents. Several countries, including Bolivia, Colombia and Venezuela have decentralized extension, placing responsibility for it at the municipal level where its links with research may be distanced.

Although some of the reasons for the lack of sustainable financial support from the public sector have been explained above, there are other reasons why the INIAs have little opportunity to exert any influence on the size of their budget. For example, most of their funding is derived from a public sector that has historically practised discriminatory policies against agriculture (Schiff and Valdez, 1992). Thus, it is not suprising that after the debt crisis of the early 1980s, the structural adjustment policies that were introduced throughout much of the region resulted in reductions in the funds available for public sector agricultural service organizations such as the INIAs.

Era of Change: 1990 to the Present

Throughout the region structural adjustment has been associated with a new policy environment, in which import substitution policies have been widely replaced by export-led growth models that have led to demands for agricultural intensification. The new policy environment, which has been widely adopted, favours trade liberalization, privatization and more decentralized decision making. It offers agriculture the opportunity to exploit comparative advantage based on climate, soil and infrastructure, and to become more competitive in both domestic and external markets. Also, new demands have been placed on the INIAs to help conserve or enhance the physical resource base, which is increasingly perceived as being in jeopardy (McMahon, 1996) particularly through soil degradation and deforestation.

The development of economic integration initiatives, such as MERCOSUR, is a natural consequence of these policy changes and is likely to lead to some reorganization of land use and production patterns. Elimination of trade barriers may help to rationalize land-use patterns regionally in terms of the agroecological potential, and offers new opportunities to the agricultural sector at the national level (Trigo, 1995). It also introduces risks, because it will reduce the scope for protecting domestic producers through the use of tariff barriers and, without new or improved technology, a country may not be able to exploit its comparative and competitive advantages.

The new era of economic opportunities based on trade liberalization has already led to spectacular growth in regional trade, of which about 40% is agricultural (Bathrick *et al.*, 1996). As a result of increased market opportunities emerging for higher value crops, an increased emphasis on postharvest handling and developments in agribusiness, the regional agricultural sector is undergoing a process of transformation. Many INIAs are only now beginning to develop the strategic and operational processes necessary to work in this enterprise environment, particularly in terms of serving smaller producers. The emphasis given in adjustment programmes to reducing the size of the public sector, to decentralization and to privatization has also altered the traditional role of the state in terms of science and technology. All INIAs have been forced to become less state-dependent and to increase their income from sales and services and from competitive grants. A number of governments are providing funds for such grants and using them as an instrument for developing research capacity in the universities and the private sector, in an effort to develop a true national agricultural research system (NARS) rather than relying on a single INIA.

A range of other strategies are being adopted by the INIAs to lessen their dependence on public funds and to improve their prospects of sustainability. Chile was the first country to adopt a market-driven approach as it moved from a state-dominated to a free market economy in 1975 (Venezian, 1992). The export of fresh products (now valued at over US$2 billion annually) was given top priority. In 1976 joint public–private sector funding was made available to create Fundacion Chile which offered competitive grants for innovative agricultural and other research in areas where Chile had comparative or unexploited advantage (Corning, 1993). Much of this research was revenue generating and Fundacion Chile has become fully sustainable. In 1981 the Ministry of Agriculture created a competitive research fund (FAI) specifically for the agricultural sector. Following this, several other competitive development funds were established including one created by the private sector for fruit research. In 1987 a law was passed permitting tax deductions for research donations to universities. Currently, about 40% of agricultural research is conducted outside of the INIA, much of it by the universities, and INIA now receives only one-third of its budget directly from the government.

In Venezuela about half of the national research budget is now implemented by the universities. This situation is associated with the severe reductions which the budget of the INIA (FONAIAP) has suffered in recent years. In the Caribbean, the University of the West Indies is actively engaged in agricultural research as are several major universities in Brazil. In much of the region, however, the research potential of the universities still remains largely underutilized and the basic and strategic research areas, where they have the resources to make a major contribution, remain neglected.

In Brazil, where several public agencies, and also the private sector, offer funds for agricultural research, EMBRAPA has followed an entrepreneurial approach capturing about half of the available private sector funds in addition to gaining about 10% of its budget from competitive grants offered by the Secretariat of Science and Technology and a further 10–15% from its sales and services (Alves, 1992). A specific unit has been established to assist in cost recovery and to expand the sale of EMBRAPA services. This type of enterprise culture also exists in Argentina, where Fundacion ArgenINTA has been created to market the products and services of the INIA (INTA), including participation in joint ventures. This foundation, like Fundacion Chile, seeks to link the science base with innovation and investment opportunities. The Uruguayan INIA has a joint venture with a malting company, and in Colombia, CORPOICA conducts contract research on behalf of the rice, cereal and other producer associations.

In Argentina, INTA was formerly funded by a tax on agricultural exports but this has now stopped. A similar export tax exists for coffee in Colombia, with part of the revenue being used for research. This method of funding involves major risks when research revenue is dependent on the price for a commodity such as coffee where world market prices are highly volatile. Colombia has recently adopted a number of other measures to build a stronger and more sustainable agricultural technology base. The INIA (ICA) became a parastatal organization (CORPOICA) with the freedom to undertake contract research and also to contract research out to other bodies. Its staff are on tenured contracts and it has to obtain part of its budget from successful tendering in a new competitive fund. Colombia has a strong group of producer organizations, some of which have been accustomed to financing research on either a voluntary basis or through an export tax (Posada, 1992). Recent legislation recycles new commodity production levies to the commodity associations which have the responsibility of managing the funds. Many of the associations are using a major part of this revenue for research. Some associations conduct their own research, others contract it to CORPOICA, CIAT and universities. Currently CORPOICA's core budget represents about half of the funds available nationally for agricultural research.

Another new piece of Colombian legislation taxes imports of soya, barley and wheat, which together provide an annual revenue of about US$6 million which is currently destined for technology transfer rather than research. In Peru, a Law of New Investments offers incentives to producers to invest in research and development. In Uruguay about 30% of the INIA budget is derived from sales and services, with the balance coming in equal parts from a tax of 0.25% of the value of production at the first point of sale (which has to be recorded legally) and a matching government contribution. INIA is managed by a four-person directorate; two government and two farmers' association nominees. The directorate offers 10% of the INIA budget for competitive funding to be used by other institutes either alone or in partnership with INIA.

In the Mexican state of Sonora there is a unique example of research funding in which a farmers' association (PATRONATO) with about 10,000 members voluntarily contributes a levy of about 0.16% of the value of their wheat production. This generates about US$400,000 annually, matched by a similar sum of public sector grants, to pay for research by the INIA (INIFAP) and by CIMMYT. This research is carried out on lands belonging to the PATRONATO. Through demand-driven research, members of the PATRONATO have increased their average wheat yields from 3000 kg ha^{-1} in 1964 to 5500 kg ha^{-1} in 1995, and established a successful new soya/wheat rotation on 250,000 ha of their lands.

The above examples illustrate that a wide range of new funding alternatives are being adopted in the LAC region to increase the likelihood of sustainability. Three measures are being pursued particularly actively, namely competitive funding, agricultural science foundations and greater private sector participation (Echeverría *et al.*, 1996a,b).

Competitive funds of various types already exist in countries such as Argentina, Brazil, Colombia and Chile and in new World Bank projects in Ecuador, Brazil and Bolivia. The Inter-American Development

Bank is about to launch a regional competitive fund to support agricultural research that has regional implications (Birdsall, 1995). This regional initiative has been created on the basis of the past achievements in terms of comparative advantage, critical mass and research results obtained through: (i) regional cooperation in the Programme of Regional Co-operation (PROCI) networks that are funded by BID and managed by the Inter-American Institute for Co-operation in Agriculture (IICA); and (ii) the various commodity-based networks of the regional IARCs.

Non-governmental research foundations supporting agricultural research, such as Fundacion Chile, FUSAGRI (Venezuela) and Fundacion Patiño (Bolivia) have existed in Latin America for a number of years. During the 1980s USAID created and supported new private sector 'foundations' in Honduras, Jamaica, Dominican Republic, Peru and Ecuador. These are discussed later in this chapter. Increased private sector involvement in agricultural research has taken place spontaneously in many countries and has increased as the public sector has opened up the role of the INIAs. Initially, the main effort was from multinational suppliers of inputs in the larger markets of Brazil, Argentina and Mexico. Private sector cereal research is already important in Brazil and Mexico. With markets expanding elsewhere and INIAs facing recurrent financial constraints, new opportunities are arising for private sector research in areas such as tissue culture and vaccine production, particularly where the technology can be appropriated. Good data are hard to obtain in this area but it has been reported that 30% of agricultural research in Ecuador is now privately funded, 38% in Colombia, 56% in Jamaica (Falconi and Elliott, 1995) and 35% in Uruguay (Indarte, 1997). More recent data for Colombia show a figure of 50% if rebates from specific commodity levies are included as producer contributions (Jimenez, 1997). Currently only a few producer associations in Colombia conduct their own research, which is contracted out, often to CORPOICA.

Many of the new initiatives described above are too recent to be evaluated in terms of their sustainability. Furthermore, even with these initiatives few of the NARS of the region have a total budget that is much more than about 0.6% of the value of the agricultural GDP and, despite considerable institutional evolution and division of funding sources, the public sector is still the dominant actor in funding and implementing research (Echeverría *et al.*, 1996a).

> The key problems confronting agricultural research in the region relate to: how the decline in funding can be reversed; how the role of the private sector can be expanded; how producers can play a larger role in defining research programmes and in supporting them; how the transfer of technology to producers can be improved; how the national research system can be made more pluralistic (i.e. become a NARS rather than

being INIA dominated); how this NARS can be positioned in the global research system; and finally, how the efficiency of resource use can be improved and how national systems with limited resources should respond to new challenges such as new information technologies and biotechnology.

(World Bank, 1996)

In the following pages these issues are discussed in the context of the lessons that can be drawn from past experience.

Lessons for the Future

Lessons for governments

Governments can draw on a number of lessons from past experiences in order to promote appropriate and sustainable agricultural research systems. It is important to record these lessons because it is a feature of the political system in many LAC countries that the tenure of the Minister of Agriculture is relatively short and the leadership of the INIA is a political appointment, which is often discontinuous (in one country the INIA recently had nine Directors-General in only 6 years). One effect of these changes is that, within the government and the INIA leadership, the institutional memory relating to research tends to be short-lived and the lessons from past experiences are frequently overlooked as new Ministers and Directors-General introduce their 'new' (and often transient) policies. However, a number of countries are now appointing the leaders of their INIAs on the basis of competitive merit. In Brazil the most senior posts in EMBRAPA will now be selected by an Administrative Council on which both the public and the private sector are represented.

The public sector has an important role to play
Throughout the LAC region, past experience has shown that governments have taken few steps to encourage agricultural research outside of the public sector INIAs but, at the same time, they have generally been unable to provide adequate funding to support the INIAs at a level whereby they can function in a stable and sustainable manner. Although governments have often prioritized small resource-poor farmers as the prime clients for research, they have seldom defined what 'new' public goods should be produced for them nor what non-public institutional alternatives and funding mechanisms are either feasible or desirable (Trigo, 1995).

In these circumstances the question can be posed as to whether the INIAs need to be sustained. The short answer to this is that there are many small farmers in the LAC region who produce mainly basic food crops for their own subsistence and local market sales. These resource-poor farmers are unlikely to be able to pay for research. Unless the public sector covers their technology needs they will slip even further back in terms of the equity goal that features prominently in most national development plans. Additionally, given that non-public research is unlikely to take place unless its potential benefits are appropriable, the prospects for the private sector taking the lead role in research on natural resource conservation and management are also limited. The necessary public good research in this area could be carried out either by the INIA or by the government contracting the work to other institutions. Given the historic investment in the INIAs and their staff, and the expertise that they offer in basic food crop research, there will be comparative advantage for most, if not all, governments in channelling much of their public good investment for agricultural research through the INIAs.

Thus, there are certain areas of research where the benefits are unlikely to be appropriable. Throughout the LAC region these areas are important for national development and it is essential that the public sector continue to play a role in financing them. In many cases the INIAs will have comparative advantage for doing this research.

INIAs require adequate public sector support

If governments support public sector research, then experience suggests that their support is likely to be of limited value unless they define and limit objectives so that they are compatible with the funds likely to be available, and they also ensure that the level of funding provided is both stable and sustainable. Clearly there is little point in arguing for a public sector support level of even 1% of the agricultural GDP when experience shows that for most LAC countries the current figure is usually of the order of about 0.4%. Likewise, there is a large volume of experience from the region to illustrate that there is little value in paying for the salaries of staff who have no operational funds with which they can conduct research.

It is also important that governments ensure that the research budget that they provide is utilized in a rational manner with at least 30% being available for operational costs. Experience in many LAC countries has shown that when operational costs fall below this level the INIAs fail to function effectively. Another important aspect of financial management is to ensure the regularity of the cash flow. Many INIAs are impeded in their work because they have to operate, sometimes for months, without staff or travel costs being paid on time.

Governments have invested heavily in training research staff but have often offered these staff such poor terms of service that throughout the last 30 years the INIAs have suffered a considerable 'brain drain' with, perhaps, over half of the agricultural scientists in the region holding a PhD leaving the public service. While this loss of staff has provided undoubted benefits to private sector and commercial farmers employing the scientists, particularly in Chile and Colombia, it represents a heavy cost for the public sector, especially if it is borrowing US$100,000 of development funds to train a person who will not contribute directly to the objectives of their training. Research does not readily lend itself to the rigidity of most public sector organizations in terms of rewarding merit. Some governments have recognized this and have attempted to overcome it by partially or fully privatizing the INIAs.

Another area where INIAs have suffered rather more in LAC than in the other regions has been in research leadership appointments being associated with political patronage. A number of INIAs are now led by scientists with experience in both research and research management. But in others the leadership changes every time a Minister changes and the appointment is made on the basis of patronage rather than merit. In some countries supplementary benefits obtained through donor projects have supplemented the salary of the leadership post and enhanced its political appeal. For government funding for an INIA to be effective, the following actions need to be implemented.

1. Public sector funding provided to the INIA must be subjected to tight financial management policies and accountability with regard to its long-term stability, regularity of disbursement and distribution between cost centres.
2. Governments must ensure that the basis for appointments to the top management of the INIA is technical skill and managerial experience rather than political patronage.
3. There is an open and competitive market for good scientists and in order to attract and retain good researchers it is necessary to ensure a career structure that offers incentives for performance. Brazil and Uruguay have been more successful than most countries in achieving this, thus demonstrating that it is feasible within the LAC context.

Public sector can stimulate alternative funding sources
Private sector support for research can always be expected to relate to whether the findings can be appropriated, hence private sector plant breeding research, for example, is likely to favour hybrid production and industry-sponsored agrochemical research to focus on crops or varieties likely to respond to the fertilizer, pesticide or herbicide under trial. In

contrast with this, the progressive elements in LAC producer organizations have recognized the importance of a broader spectrum of non-appropriable research and the rationale for producers paying for near-market research that could be of early benefit to them. Producers have, however, been accustomed to the government paying all research costs and even though this has limited the amount of research carried out, many still expect research to be a free good. This is not a satisfactory situation for developing a modern science-based production sector and several governments have, with the support of progressive producers, introduced legislation to collect levies for research funding from producers, either on a global or a crop/livestock-specific basis.

Historically producer levies have been unpopular because the revenue has often not been returned to producers. Several countries, however, have enacted legislation to ensure that the funds collected in this way are turned over to producer associations for them to use for research or other development functions.

A range of other legislative measures has been adopted to facilitate the private sector to finance research. These measures sometimes involve abolishing disincentive legislation such as that giving a monopoly in plant breeding to the INIA. Incentive legislation has included tax concessions for investments in research, or for universities; changes in seed and patent laws to encourage local investment in these areas; taxing imports from outside the regional market that compete with domestic products; and allowing INIAs to participate in joint ventures and to undertake contracts paid for by producer organizations or the private sector. Many governments have started to introduce one or more of these measures. Colombia has gone the furthest in establishing producer levies and Argentina, Brazil and Chile have pioneered measures to build a marketing role into the structure and functions of their INIAs.

Several governments are also using competitive funding as a tool for broadening the basis of their support from the single INIA to the pluralistic NARS. For such funds to be sustainable they require either regular topping up or an endowment. In Venezuela (FUSAGRI) the endowment was provided by the Shell company. The Fundacion Chile was a joint venture between the ITT company and the government. For INIA in Ecuador an endowment came directly from the government, in the Dominican Republic it was provided by the private sector, and in Honduras and Ecuador (FUNDAGRO) endowments were provided by USAID. The constitution of the development banks prevents them from subscribing directly to endowments, but in Bolivia the World Bank and USAID are working together to establish an endowment fund using USAID funds as the capital base and a World Bank loan to support operational costs during a 5 year period while the capital base grows from being invested and untouched.

These examples illustrate that there is a wide range of legislative options that governments can use as incentives to both broaden the financial base for agricultural research and to optimize the utilization of existing resources through the development of a pluralistic NARS.

Public sector science policy must be flexible

In the 1980s structural adjustment resulted in reductions in many INIA budgets. On the downside, this led to reductions in operating costs, capital spending and advanced training. Some INIAs were sufficiently flexible to use the adjustment period to modernize. For example, Mexico undertook a comprehensive policy reform framework that involved reforming research financing, priority setting, financial management, institutional linkages, technology transfer linkages, staffing and programme development (McIntire, 1995). In Chile the adjustment stimulated agricultural growth and, as a result, the demand for new technology increased. The reduction in INIA's budget also forced it to seek new sources of finance and to become more open, competitive and responsive to market forces. Without the guarantee of public funding it increased its alternative sources of revenue from 10 to over 60% of total income (Venezian and Muchnik, 1995).

Another important area where policy is undergoing change relates to natural resource conservation and management. It is increasingly being recognized that agriculture is unlikely to be internationally competitive if its natural resource base is depleted or degraded. There is little empirical evidence for setting priorities for natural resources research, but with conservation becoming an increasingly important theme globally, and with countries (e.g. Dominica and Ecuador) now investing in ecotourism, natural resource research is now a subject of interest to many governments (Kaimowitz, 1996).

Other new research theme areas will continue to arise. Two are quite topical: information technology and biotechnology. Both can involve high levels of capital expenditure and specialized training of staff in a field where public sector salaries are not usually competitive. Particular care will be needed to formulate science policies for these two areas. In both cases the optimal solution for all but the largest countries may be some form of public–private linkages or strategic alliances with IARCs or other centres of excellence internationally.

One of the problems that INIAs have faced, and in many countries still face, is the absence of a clear national policy for research which includes specified priorities. In such a situation everything becomes a priority. However, there is widespread experience within LAC to show that the past results from institutes trying to conduct research on up to 100 separate crop/livestock species to satisfy every agricultural lobby group have not been very productive (Lindarte, 1995). This situation arises because most LAC countries lack a central agency or apex body with a planning or coordinating function for agricultural research. As a result there is rarely an effective procedure for priority setting and a great deal of research is carried out, within the very broad guidelines of government policy, by scientists who decide their own priorities. There is limited producer input to this process, although this situation is starting to change as producers become more involved in research financing. Moreover, an increasing number of producers now have technical qualifications and are able to prioritize their needs.

With an increasing quantity of finance being allocated to competitive funds, it is important, from the standpoint of resource allocation, that these funds should reflect national priorities. The World Bank has encouraged countries to create or strengthen Agricultural Research Councils that would provide some independent professional guidance in priority setting within the framework of national policy. Progress to date in this area has been limited. In some countries this appears to relate to the traditional research 'culture,' with neither the government nor the INIA wishing to hand over any responsibility for research policy. This problem has been compounded by the fact that the political responsibility for agricultural research is often fragmented between the Ministries of Agriculture, Science and Technology and Education (through agricultural universities).

For research to support a modern agriculture, there needs to be a flexible and adaptable National Science and Technology Policy with specific goals and priorities for the agricultural sector. Such a policy must recognize the concepts of comparative advantage and critical mass and should establish specific research priorities that are both consistent with the available resources and offer reasonable opportunities for success. Other than in the larger countries of the region, this means that priority should be placed on conducting applied and adaptive research and on creating the capability to identify, relate to and import upstream technology in more specialized areas.

Special provisions are required for small countries
Much of the commentary and the examples quoted in this chapter relate to the middle and larger-sized LAC countries. The region also includes small Central American and Caribbean countries, which have very small research organizations, principally in the public sector. In the Central American countries these INIAs are modelled on those of the larger countries. In the British Caribbean they are usually a unit of the Ministry of Agriculture and in the French territories they are part of the French overseas research organization (CIRAD). However, within the Caribbean region, banana and sugarcane research both

nationally and regionally, have for many years been supported financially by producer associations.

It is impractical for small resource-poor countries with limited markets to generate sufficient demand to sustain domestic private sector research. Ezaguirre (1996) has suggested that small country research systems must innovate, not necessarily by producing new technology, but in the approaches that they use to fund, organize and apply these technologies. He recommended that they focus their work on the adaptation, screening and testing of foreign technology; on policy advice and coordination; and on information analysis and synthesis. Some LAC countries have been successful in doing this through the use of a network approach such as the CIP-sponsored Central American and Caribbean Potato Network (PRECODEPA) and the Caribbean Tropical Fruits Network of IICA. But, in general, the research organizations of the small countries in LAC are in a state of crisis through having programmes that are vastly overextended in terms of the limited resources available to them. Few have any form of strategy or plan that prioritizes the use of these resources.

The research organization and structure in the smaller countries of LAC should not be modelled on the larger countries but should limit the scope of their research. They should give priority to developing skills in information technology in order to be able to import from elsewhere in the region both new technology and promising germplasm for adapting and screening. This is an approach of particular relevance to very small countries that need to identify niche markets if agricultural exports are expected to feature in their economy.

Lessons for the INIAs

However much effort the State puts into providing a suitable policy framework for the INIA to work within, the ultimate success or failure of the research programme depends on the way in which the INIA generates and transfers technology. The fact that many INIAs are in a state of crisis, despite heavy past investment in providing them with resources, suggests that not all of that past investment has yielded attractive returns, and there must be lessons to be learned from INIA history.

National research cultures must change
Although a number of INIAs claim to be autonomous or privatized, their autonomy is often fictitious because their budget levels, governance and terms of service of personnel are largely under the control of the government, which sometimes seems reluctant to loosen the ties

of control. The scientists complain about this but they, too, are not always prepared to accept the responsibility, discipline and performance accountability that is expected in the private sector. Attempts to privatize the INIA were not successful in Chile (Venezian and Muchnik, 1995) and moving the INIAs out of the public sector has presented problems for both Colombia and Ecuador. There is also opposition from some INIAs to new competitive funds as these are seen as mechanisms that could reduce the share of the INIAs in the total national agricultural research budget, which they have often virtually monopolized in the past.

Many INIAs have been slow to both introduce management training and move away from the input supply orientation of the 1960s. It is only recently that the culture of many INIAs has started to focus on serving client demand rather than on supplying technology for food self-sufficiency. Changes in institutional culture have been slow and have often been resisted by senior scientists on the grounds that they may lose their scientific credibility if their research is based on demands from areas in which they do not feel completely confident, rather than in their own specific area of speciality or interest. In this context, INIA Uruguay with its producer-oriented governance has adopted an innovative and constructive approach in allocating 5% of its budget for updating and retraining senior staff.

In the future, the INIAs will no longer have a near monopoly in the field of agricultural research. Other organizations are likely to undertake research as NARS develop and evolve, and competition for funding from various sources is inevitable, particularly given the financial policies and constraints of the public sector. *In order to be sustainable in the future, the INIAs will need to adapt their traditional science-based culture to become more responsive to market demand, more impact-oriented and more entrepreneurial in generating funds for their research programmes.*

Management of research must be improved
During the last two decades there have been many reviews of the LAC INIAs, particularly by donors seeking to strengthen them. Virtually every review has been critical of the INIAs administrative, financial and research management practices. Weaknesses in planning, priority setting, monitoring, evaluation and financial procedures are widespread and are reflected in the poor public image that many INIAs have at the national level. This situation has been partially ameliorated as a result of efforts by both IICA and ISNAR to produce management training materials and run training courses. Additionally, more recent World Bank projects have provided specific resources for strengthening management. Nevertheless, a recent review by Antholt (1994) has shown

that many management deficiencies still exist, and the overall use of existing resources is often neither efficient nor effective. Change is, however, starting to take place, particularly in the Southern Cone countries where market awareness is now quite evident.

There is a sense of frustration in the donor community with respect to providing further support to the INIAs because of their poor management. Likewise, many governments and producers are sceptical about investing in them. This is reflected in governments wishing to channel funds to other research institutes through competitive funding and some producer associations starting their own research units. The Bolivian oil crop producers association (ANAPO), for example, contract a Brazilian university rather than their domestic INIA, to do their soya breeding research, and the Ecuadorian banana producers pay the Honduran Research Foundation (FHIA) to do their banana breeding research. Future investment in research in the region is likely to shift to those institutes (including non-INIAs) that manage their resources in the most cost-effective manner. This situation will be enhanced by the increasing amount of competitive research funds that are starting to emerge. *If the INIAs wish to survive in a competitive market they will need to seriously address management issues, especially priority setting, cost control, and impact and start to demonstrate that they have put modern management techniques into practice, and that they are using the resources allocated to them both efficiently and effectively.*

Linkages can offer a cost-effective way to increase resources
Given the resource limitations that are likely to exist well into the future, especially in the smaller countries, the concept of research linkages is particularly important in the region. This concept can be implemented in a number of ways. One of the most productive, that already functions extremely well, relates to linkages between national research and the IARCs. The latter have, until the recent past, focused heavily on germplasm improvement and on training national staff, particularly in the area of germplasm improvement and evaluation. This has had a major influence in the region and much of the genetic improvement and yield increase seen in food crops in recent years has been associated with INIA/IARC links.

These INIA/IARC links have been primarily associated with food crops, and have not spilled over to cash crops where market prospects have recently been enhanced by trade liberalization. Given the opposition of some of the IARC donors to research that is not perceived to relate to poverty alleviation or environmental sustainability, the national programmes will probably need to seek other partners for cash crop research if they are to establish the type of linkages for these crops that they have had for food crops with the IARCs. This appears to be feasible, and INIAs in the region have already established relationships with other overseas centres of excellence. Several countries, including a number of Caribbean islands, have joint programmes with CIRAD from France; the UK Department for International Development (DFID) is providing grants to UK universities and private organizations such as the Natural Resources Institute; the Netherlands is supporting the University of Wageningen to provide integrated pest management expertise in Bolivia; and a number of US universities supported by USAID are linking with research units, both public and private, within the region. For example, Washington State University is linked to various institutions in Chile (Bathrick *et al.*, 1996). The producer-owned Sugar Research Institute (CENICAÑA) in Colombia subscribes with others to genetic engineering studies on sugarcane by Columbia University in New York. Linkages, also known as strategic alliances have also been fully exploited by the regional IARCs, whose partners for basic research embrace a wide range of research institutes in North America, Europe, Japan and Australasia.

Apart from upstream linkages of this type, a number of linkages exist at the working level through regional networks of which a large number exist in LAC. They are usually sponsored by international organizations such as FAO, IICA, IDRC (International Development Research Centre, Canada) and the IARCs, and are of particular value to the smaller countries. There is, however, such a proliferation of networks in the region that participation in network meetings sometimes represents an excessive demand on senior management time.

A third type of linkage, which is often poorly developed, is that between research institutes in the same country. The shortage of funding has led most INIAs not to seek national partners who may be seen as competitors. Even when World Bank funds have been provided to INIAs specifically for contract research with other national organizations, their use has been limited; thus, joint INIA/university projects do exist but are not common. In Chile, Colombia and Uruguay their growth is being stimulated by competitive funds that provide incentives for this type of link. In Colombia it was difficult to develop partnership projects before the INIA was partially privatized, because public institutions were not permitted to offer or receive funds from the private sector.

The benefits of research linkages, through strategic alliances with overseas centres of excellence, through regional cooperation and through national partnerships, have already been shown to be beneficial within the region. These types of relationships can be expected to increase in the future and need to be emphasized in the research planning and resource optimization processes, and in order to capitalize on comparative advantage.

INIAs must be ready for new challenges
A major problem in the recent past was that research programmes focused on policies established in the 1950s which were not relevant to the economic climate of the 1990s. The INIAs preserved and defended their existing programmes and culture and were slow to respond to new demands. Today they face three new challenges. The first is the emphasis on sustainable agriculture and on strengthening capacity in natural resource management research. Involvement in this public good area raises a number of issues for the INIAs, who are the logical agencies to carry out the work. But, as with so many new programmes in the past, natural resource research is usually added to their portfolio without the addition of appropriate funding both to do the work and acquire additional expertise. The challenges facing the INIAs include deciding how to balance their research between production and natural resource activities in order to accommodate the latter, whether they wish to compete for this work or to see it performed by an environmental agency or to be contracted to a university, and how they will relate to a new set of stakeholders and clients (Crosson and Anderson, 1993a).

A second challenge in the 1990s is biotechnology. A growing number of LAC countries are investing in infrastructure and human resources for biotechnology programmes and adopting policies to facilitate biotechnology research and development. Such policies involve identifying relevant end-users and developing national expertise and funding sources. Because this technology is appropriable, providing that suitable legislation exists, the private sector is likely to be interested particularly in development, marketing and distribution. This potential interest raises important policy issues for the INIAs in terms of their approach to biosafety and intellectual property rights (Persley, 1990a,b; Cohen, 1994; Persley *et al.*, 1992; Doyle and Persley, 1996).

A recent survey of agricultural biotechnology in LAC (Jaffe and Infante, 1996) concluded that biotechnology is a complex and costly process that many LAC countries have embarked upon without adequate preparation. Progress is likely to be limited by the lack of long-term capital for R&D and the weakness of the existing technology infrastructure. Particular weaknesses lie in the areas of staff training policies, client identification, an overly academic orientation and the establishment of a suitable legal framework. Another common problem is the fragmentation of effort in an area where expertise is usually widely dispersed. In Brazil, Cuba, Mexico and Argentina this problem is being approached by creating special institutes for biotechnology with sufficient staff to assume national leadership in the area.

The third new challenge is information technology. Most research institutions in the LAC region are now equipped with computers.

These facilitate easy communication with overseas scientists and libraries. Most contacts, information and even statistical results are available in minutes or hours, rather than days or months. This presents challenges as to the most cost-effective way of acquiring information and communicating with other scientists. For small countries, in particular, it presents opportunities for importing technology; it also implies that a key person in a very small research institute (particularly an INIA) should be someone with advanced training in information technology.

With the passage of time further new challenges with major cost implications can be expected to emerge. Many countries have not been well prepared for the three challenges discussed above and have not always handled them well. *When exciting but costly new challenges do emerge it is important that countries do not rush into trying to resolve them but first attempt to establish the optimal framework for action.*

Technology transfer starts when research is planned
As mentioned earlier, most INIAs no longer have a direct involvement in extension but many still retain a role in the transfer of technology to intermediary or extension agencies. This role is often put aside until the research is complete, without considering from the outset whether successful research findings are likely to be adopted. In many situations this has not been the case, often because the findings required the use of resources that farmers were unable or unwilling to supply. Although technology transfer is part of the production continuum, it is not always considered as such either by governments or by those conducting research within the INIAs. This appears to be a major shortcoming in relation to investment in research in the region.

A market-oriented approach calls for a change in this situation, and producer association support is unlikely to be forthcoming in the future unless the research carried out in the region is seen to produce results that are both relevant and easy to adopt. *Technology transfer is an integral part of the research process and should be taken into account when the research is originally planned and not left until it is completed.*

Lessons for producers and private sector

Agricultural research in LAC was for many years a virtual monopoly of the INIAs, but the share of the research budget paid for by producers and the private sector is now increasing. Although this is partially involuntary, due to producer levies, it appears that there has been spontaneous growth from some parts of the private sector and that plurality is beginning to emerge from within the countries, as well as that being stimulated by donor or government action.

Public and private sector research can be interactive or complementary
Most past interactions between the public and private research sectors have taken place through contract research. There are nevertheless some good examples of true synergy, such as maize research in Guatemala (Umali, 1995), rice in Colombia and cattle in Jamaica (Falconi and Elliott, 1995). These partnerships are increasing through joint ventures between the INIAs and private companies, for example, in Uruguay and Argentina. Additionally, research contracted by producer groups is expanding and becoming more of a partnership with producer representatives having a greater say in research programming.

In the private sector, horticultural research is flourishing in Mexico and a number of small R&D companies have emerged throughout the region because well-trained scientists have left the public service and established their own research laboratories in areas such as adaptive tissue culture and vaccine production. In Brazil, with the largest INIA in the region, a significant part of maize and soya seed research is carried out by private seed production companies.

It is possible for public and private sector research to be interactive and complementary.

Producers must support research
As mentioned throughout this chapter, there is a regional trend towards producers having to pay for the costs of research on the crops/livestock that they produce. This is a measure that is not easily introduced for basic food crops and horticultural crops produced by small farmers, and it is likely that these will continue to require support from the public sector. But crops that enter the commercial market, especially those that pass through a collection point such as a mill, port, processing plant or abattoir can easily be recorded and assessed. With the pressures that structural adjustment programmes are placing on governments throughout the region, the prospects for making near-market research the responsibility of those who will benefit from its results is an attractive proposition. It serves two additional purposes. First, it encourages efficiency because producers who pay for research will seek results that will increase their productivity. Second, by making producers more involved in the research process it should help to make the latter more demand driven. From the standpoint of the INIAs it is also likely to mean that they can expect producers to want to have a voice in the management of their funds and to have a say in the way in which the INIA is

managed. With the move towards competitive funding, both of these trends can be expected to develop, especially as some competitive funds prioritize or insist on interactive research.

In the above scenario, *producers (especially those producing commercial crops) must expect to contribute to the cost of research and, in return, should expect to have a voice in determining the nature of the research programme and in evaluating its performance.*

Lessons for the international community

Past donor support to the LAC INIAs has been extensive. It has provided physical plant, equipment and postgraduate training for nearly every country. When the use of these resources was constrained by lack of operating costs, some donors also contributed these. But the end result is that, with few exceptions, the INIAs are now in a state widely described as 'crisis' (Trigo, 1995; Echeverría *et al.*, 1996a). Past donor efforts have not had the desired long-term outcome and there are clear lessons to be learned.

INIAs must be able to survive in domestic environments

The major cost in donor projects that strengthened the LAC INIAs in the 1970s and 1980s was the construction of new research infrastructure. Donors often tried to put into place the fabric of an established research organization without necessarily recognizing that sustainability requires that governments commit the resources required to maintain this fabric. Because this has not always happened, there are a number of research stations with a good physical plant which is not properly maintained, and a considerable amount of sophisticated research equipment which is out of use through lack of funds to purchase spare parts or consumables. Even more serious for the INIAs is the wastage from the investment in personnel who have studied abroad for higher degrees only to return and then resign from their posts because of poor salaries and incentives.

Most INIAs now have adequate, if not excess, physical infrastructure and to remain sustainable a number have recently reduced both physical facilities and staff numbers. Today few donors provide further support for facilities development and support for training has been reduced because system expansion has slowed down and more qualified national staff are now available. Donors also recognize that financing operating costs, particularly the topping-up of salaries, may be counterproductive in terms of sustainability.

Institutional reform is a long and difficult process that cannot be made sustainable simply by pouring in resources. It requires relating resources to a level of maintenance costs that the government can afford and is willing to sustain'

(World Bank, 1996).

The project approach may not be the best way to build sustainable institutions

Most donors work through a 'project' approach following a format that conforms to a fairly rigid model, which relates more to infrastructure creation than to institution building. Much of the recent donor support for the INIAs has also been conditioned on the preparation of a strategic or master plan and on the setting of priorities. A recent World Bank (1996) report on donor research projects has drawn attention to the fact that many of these projects have not recognized the 15–20 year time frame for institutional development nor the resource constraints (especially financial resources) faced by governments, and has pointed out that plans and priorities are often imposed from outside, with limited local input or ownership.

A case study of one donor's support for private sector, non-profit, agricultural research organizations in LAC has highlighted a situation that is relevant to a number of past donor activities in the region. The study reveals that donor support was generally supply driven, focusing on the provision of inputs to cover the cost of a technology agenda consistent with donor priorities. The donor was more preoccupied with managing and disbursing project funds than with helping the new organizations to develop a demand-driven (market-led and client-oriented) approach to provide services and products that would generate credibility and sustainability for the organization after the donor support terminated (Byrnes and Corning, 1995).

This situation characterizes another problem with the project approach in that few projects are designed to cope with the transition from donor dependence to self-sufficiency, where the national programme has to maintain the new resources and reintegrate the staff given advanced training during the project.

Although *conditionality* has become part of the agricultural research assistance process and 'conditions' pursued have become more clearly defined and monitorable during the past 15 years (Tabor and Ballantyne, 1995), it is not uncommon for the 'conditions' negotiated to be ignored once a project is implemented. This situation is often associated within the INIA having little sense of ownership of the project and by the personnel associated with developing it being replaced by new appointments by the time of implementation. On the donor side, an undue emphasis on disbursement and a reluctance to impinge on sovereignty by insisting that agreed conditions be met, has meant that much of the effort put into identifying conditions critical for the success of the project has been wasted. In past projects in the LAC region this has been particularly true for conditions relating to personnel incentives and financial commitments.

The results from the traditional donor project approach have often been disappointing. *A new approach is required that capitalizes on the lessons learned, particularly:*

- *The importance of ensuring that the government and not the donor 'owns' the new project, even though this may make project preparation slower and more expensive.*
- *The time parameters for this type of institutional development may need to be longer than for most conventional projects.*
- *The 'conditions' originally agreed to as being necessary for the success of the project are adhered to, unless rejected by the government and the donor.*

New institutional models need the support of national government
Donors have responded to the need for change in various ways. As far back as the mid-1980s, USAID stopped investing directly in the INIAs and tried to strengthen agricultural research by creating a series of private foundations to finance research and catalyse the development of a NARS. This was a step in institutional evolution seen as part of the Agency's private sector alternatives strategy. Foundations were established in the Dominican Republic, Ecuador, Honduras, Jamaica and Peru. These foundations have generally not been well linked to the rest of the NARS with the result that they compete with, rather than supplement national efforts, especially if they set their own priorities and pay higher salaries. Although the concept was attractive, foundations established without a sizeable endowment found it difficult to raise funds and have not proved to be sustainable (Sarles, 1990). The endowments provided to Ecuador and Honduras (partly from the private sector) have given foundations in these two countries an element of sustainability. USAID is currently considering providing a new endowment to Bolivia as part of a joint project with the World Bank. Byrnes and Corning (1995) have suggested that for a sustainable research organization funded by a foundation, the latter requires an income from its endowment that will cover at least 60% of the annual research expenditure. In addition to this, foundations need to develop the capacity to recover from their clients at least a portion of their operating costs.

The World Bank has also changed its project approach which, until the mid-1980s, dealt mainly with investing in physical plant, equipment and training (Pritchard, 1990). Since then, the Bank has become more interventionist, giving greater attention to research management and accountability, in terms of both financial management and research performance. Since the mid-1990s institutional sustainability has featured strongly in new Bank loans to Brazil, Colombia and Ecuador. The Bank has also attempted to assist in the development of more pluralistic NARS through promoting a range of competitive funding mechanisms and by more fully incorporating both the private sector and universities into the national research system.

In the past decade the BID has supported investment projects in agricultural technology in Argentina, Bolivia, Brazil, Chile, Ecuador, Honduras, Jamaica, Paraguay, Uruguay and Venezuela. These projects, plus support for regional and IARC research in LAC, represent a past investment of over US$1 billion. Monitoring and evaluation have been observed as a particular weakness in the INIAs, and BID has funded both IICA and ISNAR to conduct research and training in these areas. It has strongly supported the regional approach and has, for a number of years, financed three regional research networks (PROCIs) managed by IICA. BID is now sponsoring a new fund for regional research launched in mid-1997. Membership of the fund is expected to include 16 countries in the region and several donors. It is anticipated that the fund will build up an endowment of US$200 million the interest from which will support agricultural research with regional implications (Inter-American Development Bank, 1996).

Another important external player is the CGIAR, which has had greater technical impact on the INIAs than any other agency. In recent years the CGIAR has shifted its focus from germplasm development towards the more efficient use of technical inputs, combined with the progressive use of ecological and environmental perspectives, and the increased use of information technology. This change in focus has been influenced by the domestic research agenda of the donor countries and their concern to ensure that knowledge in these new programme areas should be available globally. However, many INIAs still regard germplasm development and training as high priorities and some have expressed concern about the IARCs reducing traditional activities in favour of ones that many INIAs would currently have problems in incorporating into their programmes, notwithstanding their importance.

In addition to USAID, the principal bilateral donors in the LAC region include Canada (in the Caribbean), Germany, France, Holland, Japan, Switzerland and the UK. All have tended to support the project approach, usually involving the INIA, and they often provide technical assistance specialists as a component of the project. A major strength of these bilateral programmes is that they frequently provide grants for training and technical assistance, whereas if these items are included in Bank projects they are usually loans that have to be repaid. In contrast with this, the two development banks are able to mobilize larger sums of money for improving physical facilities and procuring equipment and they also carry the influence to help governments address policy issues (World Bank, 1996). In some projects in the region the multilateral and bilateral agencies have formed partnerships to capitalize on their respective comparative advantage.

A lot of creative thinking is now being devoted to making donor investments in agricultural research in LAC more productive and sustainable. To do this it is necessary to avoid some of the errors of the past, including: creating donor dependency; fragmentation of research efforts; lack of programme continuity; failure to build up counterpart funding needs for recurrent costs to a sustainable level; poor utilization of some inputs; and inflexibility in linking research funding to policy reforms (World Bank, 1996). Many of these errors can be attributed to an inadequate dialogue between donors and governments.

There are opportunities for donors to support reform of the agricultural research systems in the LAC region, but national governments need to be more fully involved in the development of new institutional models and donor support programmes and both they and donors should give more attention to the factors that have constrained institutional sustainability in the past.

Institution building skills are a scarce donor resource
Although much has been written about the limited management skills of INIA staff, the same comment is often valid for donors. There appears to be an inadequate dialogue between donor policy-makers whose reports indicate that they are aware of the difficulties and needs of institutional development, and donor operational staff who may have to work within a 'project model' that is insufficiently flexible to recognize that institutional projects are highly people-oriented and cannot necessarily be implemented in the same way as, for example, building roads or schools.

Projects are often prepared with a sound initial technical base but this becomes distorted as proposals pass through complex donor approval procedures. Project preparation is also becoming increasingly dependent on short-term consultants, since many donors have reduced their numbers of experienced technical staff. Once a project is agreed to, its management may be assigned to staff with little technical background whose management skills lie in ensuring that disbursement schedules are met rather than institutional development goals achieved. Donor projects are sometimes managed in such a way that there is no clear line of accountability for performance within the donor agency. Tabor (1995) has pointed out that the skills and background necessary for planning and evaluating technical investments are not necessarily the same as those required for defining a framework for research system reform. Few donors have staff working on the changing role of agriculture within the existing trade liberalization context, which is likely to drive the direction of agricultural research in the next decade (Bathrick *et al.,* 1996).

The preparation and management of donor projects associated with the strengthening and reform of agricultural research is a career task that requires the use of skilled human resources and clear lines of responsibility.

Investment Strategies for the Future

Most of the lessons discussed above focus around three theme topics:

1. How to create and sustain a broad-based funding mechanism in which the efficiency of public expenditure is increased by making it more focused and effective through competition and accountability, and by creating a more favourable environment for the private sector (including producers) to invest in agricultural research.
2. How to establish and sustain a pluralistic NARS in which the principles of comparative advantage apply to the institutional development process so that the various partners such as the INIAs, universities and privately funded research organizations each play the role to which they are most suited.
3. How to increase the efficiency of resource use by the INIAs through improved priority setting mechanisms, better research management, a more entrepreneurial approach, greater accountability and increased interaction and linkages, both nationally and internationally.

This chapter has indicated that a number of measures have been adopted in the LAC region in the 1990s, which are targeted at these two goals. They include steps to:

- focus public sector funding on public good research, and on catalysing the creation of a holistic NARS, particularly through establishing the legal framework to facilitate this;
- stimulate the private sector to assume more responsibility for funding research;
- finance research on the basis of demand;
- establish competitive funds for research, open to all members of the NARS;
- improve research management and increase accountability based on performance, both technical and financial;
- assist the universities to play a more active role in a knowledge-based agriculture.

There are a number of mechanisms by which donors should be able to assist LAC governments in the further implementation of these measures. Steps that might be taken by donors include the following:

1. Assisting countries to establish the legal and policy framework for developing a NARS. This requires redefining the mandate of the INIAs, setting clear policies for public sector research, and providing appropriate incentives for other organizations to become more actively involved in research financing.

Development banks are the donor group most appropriate to engage in this role, other than in the small Caribbean territories where donor assistance is often closely linked to a single bilateral donor. The Economic Development Institute of the World Bank could play a significant role in this area with respect to the training of government policy-makers and research managers, and also Bank and donor operational staff. This chapter has illustrated that there are a number of experiences of successful change mechanisms available from the LAC region although some countries and donors are still investing in the context of an institutional framework whose current relevance may no longer be valid.

2. Stimulating the private sector to contribute to research financing through measures such as legislative incentives for investing in research or the removal of disincentives, the creation (with the support of producers) of levies giving them the responsibility for payment and management of near-market research.

This is an area that is suitable for *bilateral assistance* because it requires the type of technical expertise and training skills possessed by many donor countries that have commodity boards experienced in financing research in this manner.

3. Providing support throughout the NARS, and particularly to its universities, by fostering strategic alliances with research institutes in both the CGIAR and donor countries.

This is another area where *bilateral donors* can exploit their comparative advantage in specific technical or commodity topics. They could also play a significant role in the development of the NARS by promoting the sort of collaborative public–private–university linkages that are common in the North but rare in the South.

4. Helping to make the INIAs more entrepreneurial by improving their management, especially in the areas of finance, public relations and marketing. This entails creating the skills required to sell their products and services more effectively, to tender successfully for competitive funds and to optimize their resource use for embarking on joint ventures and strategic alliances.

This area is particularly suitable for *bilateral donor assistance* through technical assistance and training from institutions and personnel (particularly from the private sector), which have a proven record in research management.

5. Supporting mechanisms for establishing sustainable competitive funds. The establishment of such funds appears to be a lynchpin of the present strategy for developing a pluralistic NARS in a number of countries. Regional experience to date suggests that this type of mechanism will only be sustainable if the competitive fund is created with an endowment that is large enough for it to function effectively on the interest from its investments until replenishment mechanisms have been built up.

The Fundacion Chile, which is probably the best example of a sustainable competitive fund (although many of its activities are outside the agricultural sector), has certain key characteristics that relate to its success (Corning, 1993): (i) it has a clearly defined set of priority objectives; (ii) it has a rotating non-political governance in which potential users of new technology are represented; (iii) it is demand oriented so that the projects that it supports have a potentially practical benefit; (iv) grantees have to make a meaningful contribution to projects from their own resources; (v) a significant number of projects have to offer prospects for generating some form of revenue to the Fundacion so that it can sustain its resources; and (vi) it monitors and evaluates the performance of its grantees and bases future awards on past performance.

Competitive funds lacking the above characteristics have failed to be sustainable. It is important that this be borne in mind given the current enthusiasm that both governments and donors are showing for such funds.

Providing support for competitive funds offers an opportunity for the *development banks* and *bilateral donors* to work together in their support for research. The banks are constitutionally barred from investing in endowment funds but are able to offer loans and policy advice to governments establishing such funds. Bilateral donors, on the other hand, are usually able to offer government grants as well as loans. Furthermore, a number of them are interested in encouraging the type of private sector and university research that can be stimulated by such funds. The type of partnership arrangement that is being developed in Bolivia may serve as a model for this type of joint approach.

6. Another area of opportunity for joint action by *development banks* and *bilateral donors* relates to the IARCs, whose past activities have made such significant contributions to research and training in the region. The IARCs now face a number of new challenges without having sufficient resources to handle them, and also to conduct the necessary maintenance research and training that are an essential follow-up to their past LAC programmes at the national level. To maintain their effectiveness the IARCs will require progressively greater resources in the future or a redirection of their current resources.

Conclusion

In spite of the disappointing results of the last 30 years in creating sustainable research organizations in the region, the results from some of the newer strategies discussed in this chapter offer promising prospects for future change. For such change to be effective, new policies will need to go hand in hand with changes in the cultures of all of the partners in the research enterprise. Governments will need to loosen their past monopoly so that research takes place in an open market. The INIAs will need to develop a performance-based culture. Bilateral donors will need to ensure that their support is demand based and not driven by pressures of domestic supply. Multilateral donors that base loans on specific conditions for implementation will need to make more effort to ensure that conditions agreed to with governments when projects are initiated are adhered to. And INIAs and donors can both usefully improve the training and outlook of staff involved in developing and managing research institutions and projects.

There are many interactions between the parties involved in the success criteria discussed above. The chances of bringing about successful change will be enhanced if the dialogue and sharing of experience between governments, INIAs, universities, the private sector and donors can be improved. Perhaps the most important measure that can be taken towards bringing this about will be for governments to either create national apex bodies for agricultural research policy or, even better, to ensure that there is a national body for science and technology policy, including that in agriculture, which has both technical capability and political support. By doing this there will be a realistic opportunity to create a genuine national agricultural research *system*, and to link it to the emerging global network of agricultural research. Much of the global framework is already in place, and the challenge to the LAC countries is to create new institutional settings that will enable them to participate in it more fully in order to optimize the use of their own resources.

Brazil

6

Francisco J.B. Reifschneider,[1] Uma Lele[2] and Alberto D. Portugal[3]

[1]Secretariat for International Cooperation, EMBRAPA, Caixa Postal 040315, 70770-901 Brasilia DF, Brazil; [2]ESDAR, The World Bank, 1818 H Street NW, Washington, DC 20433, USA; [3]EMBRAPA, Caixa Postal 040315, 70770-901 Brasilia DF, Brazil

Introduction

Strategies for financing agricultural research in Brazil are reviewed in this chapter. The Brazilian agricultural research system (Sistema Nacional de Pesquisa Agropecuaria, SNPA) is in transition from one in which public resources have been the primary source of financing (and public institutions are virtually the exclusive source of research output and technology transfer), to a new multi-source, multi-institutional (public/private) system. The new system will allow the public sector to more sharply focus research funding and output on public goods (i.e. research whose benefits are long term and cannot be easily appropriated by individual entities, and therefore are unlikely to be funded by the private sector), generating specific solutions to regional and national problems.

Rapid changes at the global level, including advances in science, privatization of intellectual property, environmental consciousness and WTO rules in the context of expanding international trade, have made it imperative to seek diversified sources of research funding, to ensure research outputs. The challenges faced in financing this transition are discussed in this chapter. The Brazilian data are based on three key papers (Contini *et al.*, 1997; Albuquerque and Salles-Fûho, 1997; World Bank, 1997b). These papers are complemented by an interesting description on financing agricultural research in Latin America by Echeverría *et al.* (1996b).

Agricultural Research System

Brazil has about 5400 researchers, in about 50 research-related public institutions at the federal and state level (Empresa Brasileira de Pesquisa Agropecuaria, EMBRAPA, the state research and extension organizations, and the universities), the third largest agricultural research system among developing countries. China has nearly 50,000 researchers and India about 26,000. Its annual research expenditures at 0.9% of the agricultural value-added make Brazil's agricultural research system two to three times better funded than those of China or India, whose annual research expenditures stand at around 0.3–0.4% of their value-added in agriculture. Research expenditures per person engaged in agriculture are also much higher in Brazil (US$149) than either China (US$25) or India (US$15). Brazil has a higher per capita income (US$3640 in 1995), and smaller share (23%) of the national population (152 million people) engaged in agriculture than China and India (per capita incomes of US$640 and US$340 in China and India respectively, and with 74 and 64% of their national populations of 1.2 billion and 930 million reportedly engaged in agriculture). Brazil also has a larger share of its agricultural scientists trained in industrial countries. Unlike many, particularly small, developing countries, less than 5% of the Brazilian national agricultural research investments came from external loans in 1994–1996, mostly from the World Bank and the Inter-American Development Bank. Cooperation with industrial countries in recent years has been largely in the form of research partnerships.

The overall budget of the Brazilian SNPA is about three times, and the scientific staff over five times, the size of the Consultative Group on International Agricultural Research (CGIAR) system, making it a well-equipped and modern system among developing countries. EMBRAPA, which was created in 1972, was officially appointed leader of the SNPA when it was established in the early 1990s. In 1996 EMBRAPA had a full-time research staff of 2093 with an annual budget of nearly US$550 million (Fig. 6.1). The combined state research staff in 1996 amounted to 2341, with total budgets of the state research and extension systems of over US$300 million (Tables 6.1 and 6.2). The universities had an additional estimated 960 full-time research equivalent staff in 1993. It is estimated that nearly 90% of Brazil's annual agricultural research expenditures is currently in the public sector; the shares of public and private sector financing are about half and half in several industrial countries.

Human and Financial Resources

Financing and managing of a national agricultural research system poses several challenges for Brazil, because of its immense size, regional

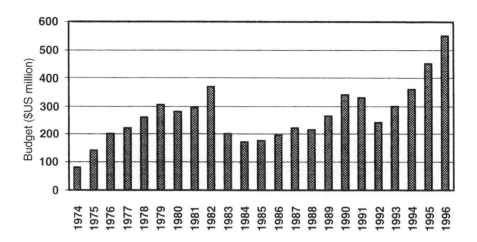

Fig. 6.1. Evolution of EMBRAPA's budget (in 1996 US$ millions).

Table 6.1. Allocation of EMBRAPA staff by region and category, 1996.

Region	Category	
	Researchers	Administrative and technical support
North	282	1035
Northeast	334	1039
South	355	1174
Southeast	434	1160
Centre-west	419	1354
Central administration and special services	269	952
Total	2093	6714

Source: EMBRAPA–SSE (1996).

diversity in natural resources, incomes and the degree of commercialization and urbanization. EMBRAPA has 37 research centres (13 ecoregional centres, 15 commodity centres and 9 thematic centres). Seven centres are located in the northeast, the area which contains the largest proportion of small farms and the majority of Brazil's rural poor population. Six EMBRAPA centres are in the north, the region that contains much of the Brazilian forestry resources and biodiversity. EMBRAPA's strength in the southern half of the country is mirrored in

Table 6.2. Allocation of the state agricultural research organization staff by state and category (1996).

	Category	
Region	Research	Administrative and technical support
Northeast	530	2063
South	551	1854
Southeast	1083	4746
Centre-west	150	604
North	27	—
Total	2341	9267

Source: EMBRAPA–SSE (1996).

the state research and extension systems. They too demonstrate great variability in their capacity to address the complex problems of agricultural competitiveness, productivity, equity and sustainability. States in the north lack state research and extension systems. Even among states with research and extension systems, there is considerable variability in terms of the size of the research budgets, human capital, salary and incentive structures, and the level of extra-treasury resources they mobilize (Table 6.3). For example, the budget of the state of Santa Catarina was US$65.5 million in 1996 compared with that of the state of Maranhao with only US$1.6 million.

Only 35% of the researchers in Rio Grande do Sul had either MS or PhD degrees, compared with nearly 100% in Mato Grosso do Sul. Less than 5% of the total resources in Alagoas was generated from extra-budgetary resources in 1996, compared with 35% in Mato Grosso, and less than 10% of the total personnel in Alagoas are researchers compared with 45% in Ceara. The average monthly salary of researchers was about US$700 in Alagoas compared with US$6000 in Rio Grande do Sul.

Research management

EMBRAPA has a well-developed system of research priority setting, planning and implementation, with 13 distinct research programmes. EMBRAPA collaborates in research with over 200 national and international institutions, public and private, including 40 national universities and many North American and European research and teaching institutions. In addition, in 1996, EMBRAPA collaborated with 93 farmers' cooperatives, 53 producer organizations and 32 NGOs on

Table 6.3. Budget and relative size of state agricultural research and state agricultural research and extension organizations (1996).

State	Budget (million Rs)	Research and extension	Relative size[a]
Maranhao	1.6	No	Small
Alagoas	1.8[b]	No	Small
Rio Grande do Norte	3.9	No	Small
Paraiba	4.9	No	Small
Ceara	8.7	No	Small
Espirito Santo	10.8	No	Small
Sergipe	11.5	Yes	Small
Tocantins	11.5	No	Small
Rio Grande do Sul	14.0	No	Medium
Rio de Janeiro	15.0	No	Medium
Mato Grosso do Sul	15.2	Yes	Medium
Parana	17.5	No	Medium
Mato Grosso	20.0	Yes	Medium
Golas	21.5[c]	Yes	Medium
Minas Gerais	22.0	No	Medium
Pernambuco	25.3	No	Medium
Bahia	30.7[d]	Yes	Large
Sao Paulo	48.6	No	Large
Santa Catarina	65.5[e]	Yes	Large

[a] Cluster analysis was used to calculate the relative size.
[b] 1995 budget: 1 Real=US$.
[c] Executed until October 1996; forecast: R$45 million.
[d] Executed until October 1996; forecast: R$38 million.
[e] Executed until October 1996; forecast: R$85 million.

issues related to the generation and transfer of technology. EMBRAPA is helping to make research outputs more responsive to the real needs of its diverse clients. The organization is accountable to Brazilian society and must generate solutions to specific, regional and national problems related to agriculture and the use of the rural environment.

Returns on research

According to the World Bank's Operations Evaluation Department, returns on Brazil's agricultural research investments have been well documented. Most such estimates focus on traditional commodity-oriented agricultural research programmes and show returns ranging from 22 to 197%. Many important studies on rates of return to research have been done. High rates of return are illustrated by Brazil's leadership in research on soybeans (179%), maize (191%) (which is discussed later), citrus (78% and higher), other fruits and vegetables, and acid soils, to name only a few.

Restructuring the SNPA

Despite its many accomplishments, some weaknesses of the SNPA have become evident in recent years. Organizational modernization indices (Table 6.4) for the state organizations indicate a strong variability among states (e.g. Alagoas 15.6 vs. Rio Grande do Sul 58.9), a significant weakness of the SNPA. The dominant public research system was becoming overly bureaucratic and was not sufficiently responsive to the rapidly changing external and internal environment. Although EMBRAPA's budget increased almost eightfold in real terms in a little over two decades since its establishment in the early 1970s, there have been considerable year-to-year variations in the financial support to EMBRAPA (Fig. 6.1). The fiscal decentralization introduced by the government in 1988 restricts the mobilization of resources by EMBRAPA to the weaker state research systems, which in 1996 amounted to US$9 million in staff support alone. Financing of the state agricultural research and extension systems has become weaker

Table 6.4. Organizational modernization indices[a] for the state agricultural research organizations.

State	Index	State	Index
Alagoas	15.6	Sao Paulo	41.8
Maranhao	22.5	Mato Grosso do Sul	46.8
Goias	27.9	Bahia	47.4
Sergipe	29.0	Minas Gerais	48.5
Paraiba	29.4	Santa Catarina	50.3
Mato Grosso	31.3	Rio Grande do Norte	57.3
Tocantins	34.2	Parana	57.4
Ceara	34.6	Pernambuco	55.6
Espirito Santo	36.1	Rio Grande do Sul	58.9
Rio de Janeiro	4.16		

Source: EMBRAPA.
[a] The index considers the presence and complexity of research planning, programming and evaluation; availability and type of computer equipment; generation of own resources (extra state treasury); contracted out research; participation in scientific events; researcher age distribution; and others. The methodology used to build the indices was based on the work of Kageyama and Rehder (1993).

and more unpredictable in the past decade, with greater pressure on them to mobilize more non-budgetary resources.

As with the private sector where increased research investment is necessary to make research more responsive to their specific needs, there is also a need for sufficient and stable support of public sector research to meet long-term national research goals. Several countries offer a variety of *institutional* models of diversification of research financing and output. These models range from budgetary support and block grants in the USA to commodity-based financing and competitive grants in Australia, or a blend of the two in Malaysia. With a customer-driven approach typical of modernized agricultural sectors, these countries display five characteristics of diversification, leading to many different sources of funding and technology for users, which provide important information for the design of new strategies for resource mobilization in Brazil, namely:

1. The share of agricultural research in agricultural GDP increases, typically from less than 1% in developing countries to between 2 and 4% as in Canada, the USA, Australia, meaning substantially greater investment in research, technology development and transfer relative to developing countries in absolute and relative terms.
2. The *relative* (not absolute) share of the public sector declines over time, with the public sector increasingly focusing on the 'quintessential public goods research', i.e. research for which the benefits are long term, broadly derived and difficult to capture for the private sector.
3. The share of private sector agricultural research, technology development and transfer increases relative to that of the public sector.
4. The role of universities increases *vis-à-vis* that of public sector research systems.
5. The role of the local and state research and technology transfer systems increases in applied and adaptive activities relative to that of the federal/central government, with the latter playing a more strategic, catalytic role in stimulating research in the national research system.

Strategies for Resource Use and Mobilization in Brazil

Brazil is aiming to achieve a major transition in a short time-frame and regulations covering intellectual property rights (IPR) and plant variety protection (PVP) for the private and public sectors are in place. Private sector response will have significant implications for the sources of future research finance and outputs. These in turn will have profound implications for human health and the state of natural resources, growth of input industries, type of agricultural production, employment and international trade.

Key strategies

The Government of Brazil recognizes development of the agricultural research system as a key element of the agricultural sector development. Its new research strategy is reflected in EMBRAPA's general strategy for 1995–1998, which includes:

- performance-based allocation of resources to EMBRAPA's research centres, central units and programmes;
- closer collaboration with other institutions in the SNPA;
- establishment of new mechanisms for collaboration with, and the promotion of, the private sector resource mobilization, so that EMBRAPA can focus on the more strategic, long-term and inter-disciplinary research;
- increasing linkages with other national and international centres undertaking cutting-edge research;
- establishment of stronger downstream linkages with farmers' organizations, state research and extension services and other users of research outputs.

As a federal agency linked to the Ministry of Agriculture and Supply, EMBRAPA's social responsibilities are fully aligned with the Government of Brazil's diverse development programmes in science and technology, rural development and rural welfare. In the years ahead, EMBRAPA and SNPA must achieve high-quality outputs that will have a positive impact on the welfare of the Brazilian people, and which must be positively viewed by its urban and rural clients. EMBRAPA and SNPA must be accountable, competitive, energetic and willing to develop productive partnerships, all with the goal of improving the quality of life of all Brazilians.

Building partnerships

Partnerships between EMBRAPA, Brazil's other national agricultural research institutions and the private sector are being promoted through the recent establishment of a competitive grants programme operated by EMBRAPA. The scheme has wide participation of SNPA members in its management, and is supported by a World Bank loan for the Agricultural Technology Development Project. The project also provides investments in institutional and human development in high priority areas for research and technology transfer, while increasing the ability of the weaker institutions of the SNPA to compete effectively for funds.

Using new partnership modalities, Brazil proposes to diversify sources of funding for SNPA from the country's current heavy reliance on the publicly funded, and predominantly public sector executed, research carried out by EMBRAPA, to a more diversified and integrated system of agricultural research, technology development and transfer. EMBRAPA, as the designated leader of the SNPA, is playing a catalytic role in this transition which is intended to do the following:

- Increase sources of financing, from the state and local governments, the private sector, farmers' organizations and commodity and environmental groups.
- Develop new mechanisms for financing agricultural R&D such as check-off (levy) and matching fund programmes.
- Increase the diversity of R&D actors (universities, international and advanced country research and teaching institutions, international and domestic private sector, state research and extension organizations, farmers' organizations and NGOs).
- Increase the role of rural and urban clients in the definition of R&D priorities and their implementation, thereby increasing the relevance of research.
- Help EMBRAPA to reorient its current structure and mode of operation to address issues of decentralization and diversification of the SNPA.

Innovating public–private sector alliance: Brazil's maize seed industry
To date the private sector has been more active in commercializing research results, as illustrated by maize seed production, than in undertaking new research, perhaps as a result of a combination of factors such as inadequately trained personnel in IPR issues. This has led to limited capacity of the publicly engaged scientists to negotiate and implement contracts with the private sector, and the lack of private sector confidence in the protection of IPR in the absence of a full legislative and implementing capacity. Despite this, EMBRAPA's revenues from the sale of products, processes and services amounted to 7% of its budget in the mid-1990s compared with an average of about 14% in the US universities. However, major US universities also show considerable variation in their ability to raise non-government funds (Table 6.5).

Maize seed production
The growing importance of intellectual property makes the relative roles of the public and private sectors in financing research and development activities particularly interesting. As the CIMMYT report

Table 6.5. Financing sources for selected US universities (1994).

Universities	Investments in S&T (US$ million)					
	Government	Industry	Institutions	Others	Total	Industry (% of total)
Johns Hopkins	712	10	28	34	784	1.3
Michigan	271	27	97	36	431	6.2
Wisconsin (Madison)	286	14	52	41	393	3.5
MIT	272	56	9	27	364	15.3
Texas A&M	216	29	101	10	356	8.1
Washington	288	33	16	7	344	9.6
UCLA (San Diego)	268	10	23	31	332	3.0
Stanford	269	15	12	23	319	4.7
Minnesota	227	24	47	20	318	7.5
Cornell	202	—	66	28	296	—
UCLA (Berkeley)	191	13	68	18	290	4.4
Harvard	190	10	16	63	279	3.5
Columbia	205	2	6	23	236	0.8
Caltech	113	5	8	2	128	3.9
New Mexico	55	4	20	11	90	4.4

Source: NSF/SRS: Survey of Science and Engineering Expenditures at University, National Science Foundation, Washington, DC.

of the Brazilian maize seed industry described (Lopez-Pereira and Filippello, 1994), Brazil began to innovate on public/private partnerships in a significant way at least a decade ago. In 1987 Brazil released the first of a series of outstanding double-cross hybrids, noted for their tolerance to acid soils, their wide adaptation and superior yields. The first of these hybrids, BR-201, was developed by the national maize and sorghum research centre (CNPMS), part of EMBRAPA. At that time, maize area was expanding in the cerrados region of central Brazil. Although the private sector was not very active in seed production in the cerrados, EMBRAPA proposed that private companies begin distributing commercial seed of its maize hybrids there, starting with BR-201. The companies would purchase basic seed (the single-crosses) for producing commercial seed of BR-201 under the technical supervision of EMBRAPA.

Although large companies were then not interested in seed production, 17 small companies were interested in the programme, and the first commercial seed of BR-201 was produced in 1987 for sale in 1988. A modest 900 t of seed was sold that year, which represented less than 1% of the total maize seed market. Problems with quality control led the companies to form an association called UNIMILHO to maintain high standards, coordinate basic seed purchases from EMBRAPA and promote BR-201. These changes turned the programme around. By 1993, 18,000 t of BR-201 seed were sold, representing 17% of the hybrid seed market. The UNIMILHO group, comprising 27 companies in 1993, is now a large maize seed producer in Brazil and BR-201 a widely used hybrid. Competition in the private seed sector intensified as UNIMILHO became more important, resulting in more options for maize farmers and lower prices for hybrid seed.

EMBRAPA also benefited. Contracts with the UNIMILHO group raised revenues from the sale of parent seed and royalties collected on gross seed sales. However, some large seed companies object to the fact that EMBRAPA does not reveal which inbred lines constitute BR-201 or make them publicly available, arguing that the inbred lines were developed using public funds and should be freely available to everyone. Despite this controversy, EMBRAPA and UNIMILHO have developed a special breeding project within the CNPMS to develop hybrids to replace BR-201. Two are already on the market (small amounts were sold in 1993), and two others (BR205/206, with tropical adaptation) have been released subsequently. In addition to increasing the sources of supply of quality seed through increased competition, another benefit has been the revenues generated by EMBRAPA that in 1995 amounted to US$2.58 million from the sale of lines. Some of this (US$1.42 million) went to the maize and sorghum centre, constituting 80% of the centre's expenditures in 1995.

Improving Research Efficiency

Improving allocation of research resources and increasing the efficiency of the allocated resources presents a challenge. The questions involve relative allocation of resources between additional domestic research, its allocation among alternative programmes and capturing spillovers of external research. As an example, in Brazil, wheat programmes make considerable use of CIMMYT germplasm, either through direct CIMMYT transfers or through a CIMMYT parent used to produce an adapted variety. It has been argued that the average size of the wheat breeding research programme budget per ton of wheat produced in developing countries is nearly twice as large as that of industrial countries. This may mean that, in some cases, Brazil might be devoting more resources to breeding when it could be spending on testing research outputs from other groups for crops such as wheat, where international transferability of research is still large.

A study by EMBRAPA of its 37 research centres is beginning to measure the issue of the efficiency of research programmes quantitatively among research units within the country. Such research has potential and probably needs to be applied at the international level as well. Using data envelopment assessment (DEA) methodology, they have measured research centre performance of EMBRAPA's centres quantitatively, measures which EMBRAPA is using for funding allocations among the centres in subsequent years. The EMBRAPA effort, now being applied to institutions of two other Latin American countries, offers a common framework, but it still has some limitations which are being tackled by EMBRAPA. These include its fixed coefficient linear programming approach, the development of standard measures for comparisons of quite different research programmes with their different input needs and outputs and the weights used for aggregation. A follow-up study to measure *quantitatively* research performance across countries is being proposed by EMBRAPA jointly with international agencies such as the World Bank and ISNAR. A sound *quantitative* approach, which addresses these methodological weaknesses, might also be applied to the research and technology functions of the private sector, to assess the relative costs and benefits of conducting different types of R&D activities among alternative producers of technology.

Table 6.6. Potential financiers of research projects by programme.

Programmes	Government	Development agencies	Private sector	Private/public partnership
Natural resources	xxx	xx	—	x
Genetic resources	xxx	x	—	xx
Biotechnology	xx	—	x	xxx
Grains	xx	—	xx	xxx
Fruits and vegetables	xx	x	xx	xx
Animal production	x	x	xx	x
Forest production	xx	x	xx	x
Family agriculture	xxx	x	—	xxx
Postharvest and agroindustry	x	—	xxx	x
Environmental quality	xxx	x	xxx	x
Agriculture automation	x	—	—	xx
Development	x	xx	—	—
Rural and regional information in P&D	xx	x	—	x

xxx, strong action; xx, medium action; x, weak action; and —, no action.

Challenges Ahead

There are a number of new technologies that should be jointly developed with the private sector to accelerate their uptake by users. By offering greater protection for intellectual property, the new IPR and PVP legislation opens many more opportunities in this regard. Although the opening of the market to the multinational corporations in Brazil has considerably increased competition, it also offers the prospects of increased concentration of a few large firms in the market which might reduce competition.

What policies should EMBRAPA and the SNPA adopt to ensure greater and more stable funding by the state research and extension systems for applied and adaptive research, so that EMBRAPA can move upstream to conduct more strategic research? The various types of research activities where the private sector is most likely to be active are outlined in Table 6.6. These include improvement of genetic resources, biotechnology, grains, fruits and vegetables and postharvest technologies. Natural resources, family agriculture and environmental quality are areas left mainly to the public sector. These are some of the issues the Brazilian SNPA, under EMBRAPA's coordination, has been focusing on in stimulating a transformation to achieve more diversified sources of financing of agricultural R&D in a sustainable and diversified way for the benefit of Brazilian society.

Research Investment Strategies in Asia and the Pacific: Lessons Learned

Australia and New Zealand

7

E.F. Henzell,[1] M.C. Crawley,[2] R.W.M. Johnson[3] and E.S. Wallis[4]

[1]182 Dewar Terrace, Corinda, Queensland 4075, Australia; [2]Foundation for Research, Science and Technology, New Zealand; [3]New Zealand Ministry of Agriculture (MAF), Wellington, New Zealand; [4]Sugar Research and Development Corporation, PO Box 12050 Elizabeth Street, Brisbane, Queensland 4002, Australia

Introduction

The changes in research investment strategies that have occurred in Australia and New Zealand (ANZ) since 1975 have to be interpreted in the light of history. Many of the structural problems of ANZ agricultural research which have been addressed in the last 25 years had their origins before World War II, and some date from decisions taken in colonial times.

The early development of agriculture in ANZ was entirely the result of technology transfer from elsewhere. Prior to European settlement of Australia in 1788, the country was occupied by hunters and food gatherers who practised no cultivation, not even a garden culture. They had lived in Australia for at least 40,000 years. The subsequent economic history of Australian farming is a record of the difficulties encountered in attempting to establish a European type of agriculture in a continent that climatically and economically differed from Europe and whose principal markets were 12,000 miles away.

Things were not much different in New Zealand, where European settlement commenced a decade or two later. The Maoris had arrived 600–800 years before the Europeans and had developed a limited range of root and tuber crops. As in Australia, the European settlers in New Zealand developed by trial and error a system of agriculture that suited the physical and economic environments of the country.

It is the institutional arrangements for agricultural and natural resources research (termed agricultural research for brevity) in ANZ that may be of most interest to other countries. The origins of the agricultural research institutions that operated in 1975, and the changes that have occurred since then, are described below. It is chiefly in the period since 1975 that the two countries developed some novel institutions for financing and conducting agricultural research.

Agricultural Research Institutions in 1975

Three main types of agricultural research institution existed in ANZ in 1975. The first group comprised the departments of agriculture that dated from colonial times. (The Australian colonies became the federated states in 1901 and New Zealand attained full self-government in 1907.) The second category consisted of the New Zealand Department of Scientific and Industrial Research (DSIR) and the Australian Council for Scientific and Industrial Research, established in the 1920s. The third type of research institution was created when the agricultural and veterinary sciences, and later environmental studies, were included in ANZ universities. Mechanisms for financing agricultural research with funds provided by the rural industries of ANZ were also well established by 1975.

Departments of agriculture and agricultural colleges

Agricultural research, development and extension began in ANZ with the establishment of departments of agriculture, agricultural colleges and experimental farms or stations during 1872–1898. The New Zealand department was founded in 1892 (Nightingale, 1992), but Lincoln College had been established by farmers in 1878. The second NZ agricultural college, Massey, was not created until 1926. In Australia, by 1907, there were five agricultural colleges and about 30 experimental farms being run by state agriculture and stock departments.

The Australian state departments grew relatively rapidly with a total staff of more than 2200 in 1920, whereas there were 502 in the NZ department by 1924. The 1920s and 1930s were financially difficult for the departments of agriculture in both countries, and it was not until after World War II that they again grew rapidly. By 1975, the Australian state departments had a total R&D staff of about 3900, with about one-third of them professional (senior) scientists. The ratio of well-trained researchers had been small in the 1920s. For example, the Queensland Department of Agriculture and Stock had only about ten professionally qualified staff members out of a total staff of 515 in 1925.

The original group of Australian agricultural colleges, which were set up by departments of agriculture, concentrated mainly on education

(Black, 1976). Their contributions to research and even extension were quite limited until recently. The land-grant college model of combined teaching and research campuses was certainly considered in both Australia and New Zealand but essentially rejected. It may have been beyond the resources of ANZ at the time. In 1890 the total population, including indigenous people, was only about 3.3 million in Australia and 0.7 million in New Zealand. However, it may have been rejected from lack of vision about the value of conducting research in colleges. By 1975, these original Australian colleges had upgraded their curricula and had begun to award degrees in competition with the universities. The major changes since 1975 have been their continued conversion to degree-granting institutions and, since 1988, their merging with nearby universities.

In the New Zealand Department of Agriculture little research was achieved at first, because of a lack of qualified staff and the burden of diagnostic work and quality control (Nightingale, 1992). The NZ department had the added distraction from 1926 of a disagreement with DSIR as to who should conduct plant research; DSIR was given this responsibility in 1936. Nevertheless, research grew to account for more than a fifth of the departmental budget in the 1960s, and by 1975 the department, now the Ministry of Agriculture and Fisheries (MAF), had a total staff of over 5000 and the Agricultural Research Division had 260 scientists and 540 technicians.

DSIR and CSIR(O)

During World War I, the British, confronted by evidence of German superiority in manufacturing and chemicals, created an advisory committee to foster scientific and industrial research; then a Department of Scientific and Industrial Research, headed by Sir Frank Heath, was established in 1916. This development was noted immediately in ANZ and led to the foundation in 1926 of organizations that had a major impact on agricultural research in the two countries.

The establishment of the Department of Scientific and Industrial Research (DSIR) in New Zealand followed a report by Sir Frank Heath recommending the formation of a national science organization that would promote and organize research for the benefit of primary and secondary industries. DSIR's research encompassed a broader range than that of the Department of Agriculture, covering industrial and natural resources research as well as that for the agricultural industries, and the stated intention was not to trespass on the existing functions of the department (Nightingale, 1992). By 1975, DSIR staff conducting agricultural and natural resources research totalled 1195 with a budget of NZ$12.1 million (US$8.3 million). In addition to undertaking its own research, DSIR set up and cofunded industry institutes or associations for fertilizer and postharvest processing research (see below).

The foundation in 1926 of the federally funded Council for Scientific and Industrial Research (CSIR) in Australia followed the setting up in the period 1916–1926 of firstly an Advisory Council of Science and Industry and then an Institute of Science and Industry. Their performance had been disappointing for a variety of reasons, including lack of funds (Schedvin, 1987). The stage was set also by frustration that had built up by the 1920s with the inadequacies of scientific research in the Australian universities and in the state departments of agriculture. However, according to Schedvin, the final trigger for the reorganization of the institute and the formation of the CSIR was a change in political relationship between Australia and the UK, aimed at strengthening economic and other links within the British Empire.

Despite financial problems during the depression, the number of full-time research scientists in CSIR grew from about 27 in 1928 to 174 on the eve of World War II. The organization, renamed CSIRO in 1949, achieved a number of notable successes in applied science for agriculture in its first 50 years, and by 1975 had about 2360 senior scientists, with 40–50% of its total budget devoted to agricultural research, using the broad definition of this chapter. CSIR/CSIRO, hereafter referred to as CSIRO, was especially successful in its research for the pastoral industries, which prior to 1926 had been largely bypassed by the state agriculture departments in favour of crops and dairying. This was also the case in New Zealand prior to 1926, where the Department of Agriculture had been focusing on quality control and farm diversification.

There is little doubt that the efforts of DSIR and CSIRO did much to lift the standard of performance of agricultural research in ANZ, and to establish a high international standing for the two countries in this field (Atkinson, 1976; Schedvin, 1987), which persists to the present day. The downside of this high regard for research was that, in Australia at least, the departments of agriculture and universities sought to emulate CSIRO as soon as additional research resources became available from the 1950s onwards, leading to structural problems which became increasingly apparent after 1975.

Universities

Although universities were established in Sydney and Melbourne in the 1850s (Black, 1976), professors of agricultural and veterinary sciences were not appointed there for more than 50 years. The six state universities founded in Australia prior to 1939 were all strongly influenced by the Oxbridge tradition. The applied natural and social sciences were admitted grudgingly and slowly. The longest delay in admitting agriculture appears to have occurred at the University of Tasmania (founded 1890; agriculture admitted 1962), but there was a long delay also at Adelaide (1874; 1924) and a significant one in Queensland (1911; 1927). The only exception was the University of Western Australia. However, when 12 new Australian universities were founded between 1945 and 1975, the immediate inclusion of faculties of agriculture or veterinary science or natural resources and environment (their titles varied considerably) became the norm.

The development of agricultural studies in the New Zealand universities took a different course. None of the four colleges of the University of New Zealand founded last century, Auckland, Canterbury, Otago and Victoria (Wellington), included agricultural science, and they still do not. However, they provided tertiary-level training in the sciences underlying agriculture and later in natural resource management and environmental studies. Agricultural and related graduate education developed at Lincoln College, which instituted a degree in agriculture in 1896. It became part of the University of Canterbury in 1962 and a full university in 1990. Massey Agricultural College was founded to provide both practical and graduate courses. It became an independent university in the 1960s.

Until the 1920s the Australian universities played little part in agricultural research, consistent with teaching institutions that operated in the liberal tradition, where 'writing and research were regarded as the private occupations of the exceptionally energetic or the eccentric' (Schedvin, 1987, p. 12). The Waite Institute, University of Adelaide, and later the Institute of Agriculture, University of Western Australia, played leading roles in changing that perception in Australia. But even in 1939, the Australian universities were not part of the mainstream of Australian society. In New Zealand this separation was manifested in a marked lack of cooperation between the departments and universities in the pre-war period (Journeaux and Stephens, 1997). With rapid expansion of the number of well-qualified researchers in ANZ universities after World War II, that changed completely.

Rural Industry Research Funds (RIRF) and institutes; university research grants

DSIR was instrumental in setting up and cofunding industry research associations and institutes in New Zealand for research on manufactured agricultural inputs such as fertilizers and on postharvest processing of dairy, meat, wool, leather and forest products. Industry contributed originally on a dollar for dollar basis, though later the formula evolved to two dollars from industry to one from DSIR for the larger associations. In 1974–1975, DSIR industry grants for agricultural purposes totalled just over NZ$1.5 million (US$1.0 million), which translates to about NZ$3.0 million (US$2 million) total on the 1:1 assumption, and compares to an expenditure of NZ$24.9 million (US$17.1 million) on agricultural research that year by DSIR plus MAF.

Industry research funding on a statutory basis in Australia dates from 1901 for sugarcane; the next large levy fund to be established was that of the Australian Wool Board in 1936. In 1973–1974, 8% of total rural research expenditure in Australia was funded by levies on agricultural commodities (16% including the matching government funds). The major proportion went to CSIRO, largely for wool research. Australia had also established industry institutes, in a similar way to New Zealand, for bread (1947), sugarmilling (1949) and wine (1955) research.

Contestable government funding of university research in Australia began with the establishment of the Australian Research Grants Committee (ARGC) in 1965. Grants were allocated for pure basic research (using the Australian Bureau of Statistics taxonomy of pure basic, strategic basic and applied research, and experimental development) in universities only. In the New Zealand universities, research was funded from their block votes and some limited special grants until they were given access to the Public Good Science Fund and the Marsden Fund in the 1990s.

The situation in 1975

The first and perhaps the only comprehensive review of Australian agricultural research was that by the Industries Assistance Commission in the mid-1970s. Their survey indicated that the total investment in R, D & E in 1973–1974 was about Au$165 million (US$113 million) with Au$15–20 million (US$10–15 million) for extension. Forestry, fisheries and natural resources research were included in the survey, but natural resources research expenditure was known to have been underestimated by the respondents. About 8% of the total was by the higher education sector and 7% by business enterprises (cf. a figure of about 56% by the US private sector). CSIRO and the state departments of agriculture each accounted for approximately 40% of the total expenditure. About a quarter of CSIRO's effort was on transport, handling and processing of products beyond the farmgate.

Investment in agricultural production research in Australia (Fig. 7.1), totalled over the three main categories of government-funded agency, the state departments, CSIRO and the universities, had grown

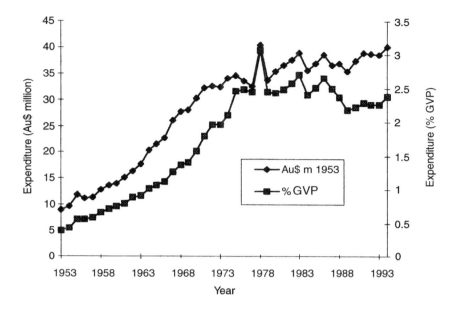

Fig. 7.1. Real and relative expenditures on agricultural production research in Australia. For departments of agriculture, CSIRO and universities only; forestry and fisheries not included. Research intensity expressed in terms of gross value of production. Source: Mullen *et al.*, 1996.

rapidly in real and relative terms from 1953 to about 1970. The universities and CSIRO grew slightly faster than the state departments. These statistics conceal the substantial change that took place in the kind of research being done by the departments. By 1975 they had developed a considerable capacity for strategic research. In contrast, the universities found themselves in fierce competition for ARGC funding and often dependent on industry and other special funds for their operational costs, which forced them towards the applied end of the agricultural research spectrum.

A particular feature of the Australian R&D scene was the relative neglect of the social sciences and of public policy advice to farmers. CSIRO's remit did not include economics, and it was not until the early 1960s that economic training gained an appropriate place in the curricula of Australian universities and agricultural colleges. In 1976–1977, less than 2% of departmental researchers were working in the social sciences and humanities. The figure was not much higher in the universities and colleges.

In New Zealand DSIR experimented briefly with a social science bureau in 1937–1940, but it failed to win political support. On the other hand, the Department of Agriculture launched economic surveys in the 1920s, and had well-developed advisory services in place by the mid-1930s (Journeaux and Stephens, 1997). In the postwar period up to 1975, DSIR and the Department of Agriculture were increasingly involved in economic surveys, combined field days and other joint activities. Business (commercial) sector investment in NZ agricultural research in the 1970s was not a large component of the total. In 1975, the government invested NZ$12.8 million (US$8.8 million) in MAF and NZ$12.1 million (US$8.3 million) in DSIR agricultural research, plus the NZ$1.5 million (US$1.03 million) that DSIR contributed to the research associations (New Zealand, 1976). In addition, MAF was given NZ$10 million (US$6.8 million) for extension. After 1945, government funding dominated agricultural research in New Zealand and expanded at a rate greater than the growth of national income.

Changes Since 1975

Resources for agricultural research

Australia
Although investment in agricultural production research grew more slowly after 1970 (Fig. 7.1), by the 1980s the national research intensity was relatively high by the standards of industrial countries. Growth occurred mostly in the universities, where expenditure on agricultural research increased by nearly half in real terms from 1978 to 1992, and in the state departments. Expenditure by the business sector on production research remained extremely low. It was about 5% of the total in 1990–1991 and about 9% in 1994–1995. Expenditure by CSIRO peaked in the early 1980s and is now in real terms about what it was in 1975 (calculated from Mullen *et al.*, 1996).

Reliable time series data are not available for expenditure on processing research, which in the Australian statistics is a subset of manufacturing, or on natural resources research related to agriculture. However, it seems certain that there has been significant real growth in both categories over the past 10 years and it is very likely that both are higher now in real terms than they were in 1975. About 60% of the rural-based manufacturing (processing) expenditure in 1990–1991 was by the private sector and the proportion has grown since.

The latest year for which comprehensive statistics are available for all three components of Australian agricultural research, production, processing and natural resources and all sectors, is 1990–1991. Total research expenditures then were in the ratio 50:36:14 for the state departments, CSIRO and universities, respectively. But 14% greatly underestimates the potential capacity of the Australian universities to conduct agricultural research. In 1990–1991, higher education institutions accounted for 22% of the total human resources, including postgraduate research students, engaged in agricultural research in the non-business sectors in Australia. In addition, there were two or three times as many university people engaged in other research in sciences and technologies which can be used in agricultural research.

The proportion of all agricultural research performed by the Australian private sector in 1990–1991 was about 17%, low by the standards of other industrial economies. With the inclusion of farmer levies, about 23% was paid for by the non-government sector. The Rural Industry Research Funds (RIRFs), levies plus matching dollars, funded about 19% of production research in the three main government agencies in 1991 (Mullen *et al.,* 1996) and about 20% of all Australian agricultural research. In addition, the RIRFs exerted leverage on up to twice as much institutional core funding. RIRF data for the 1990s indicate that the universities, the private sector and the industry research institutes receive a more than pro rata share of RIRF resources, compared with their total research expenditures.

New Zealand

During 1975–1994, government expenditure on agricultural and natural resources research in real terms peaked in the early 1980s and then declined a little (Table 7.1). The sector's share of the government's total science outlay declined from 65% to 61%. Over the period, there was a trend for the proportion invested in environmental research to increase, so the real fall was chiefly in the agricultural component.

Table 7.1. Research expenditure by New Zealand government organizations, 1975–1994.

	1974–1975	1977–1978	1982–1983	1986–1987	1993–1994
Subject allocation (%)					
Agriculture	68	65	63	65	60
Environment	32	35	37	35	40
Funding (NZ$m)					
(US$m)					
(1993–1994 prices)					
Agriculture	134.2(86.5)	133.7(86.2)	149.1(96.2)	132.5(85.5)	127.0(81.9)
Environment	63.3(40.8)	72.0(46.5)	85.9(55.4)	70.4(45.4)	83.2(53.7)

Sources: New Zealand, 1976, 1977, 1982, 1988, 1989; MORST, 1996. 'Agriculture' includes agricultural processing research. 'Environment' includes environmental protection, geological structure, land use, flora and fauna, marine and fresh water, climate and space research. Fisheries not included.

In 1994, this government investment was 62% of all expenditure on agricultural and natural resources research (NZ$336 million, US$231 million). Of the balance, 26% was invested by private entities and 12% by the universities. The greatest part of all business investment was in primary product processing research (accounting for 68% of all research expenditure in that category) with only 12% in production research and 5% in natural resources research (MORST, 1996). The greatest part of university investment was in the social sciences (65%) compared with 12% in production research and 6% in processing research (MORST, 1996). Although government was responsible for 42% of all science expenditure, it was responsible for 55% of all science funding (60% if universities were included).

Research organizations

Since the changes that occurred in New Zealand agricultural research have been far more fundamental than those in Australia, it is convenient to begin with the New Zealand experience.

New Zealand

New Zealand entered a period of major economic and institutional reforms in 1984. These reforms involved the removal of government influence (including that of direct funding), the separation of policy advice from policy delivery, enhanced government management of funding mechanisms and a desire of government to move to a fiscal surplus. Reforms of the science sector were extensive, leading initially to the establishment in 1989 of a contestable Public Good Science Fund (PGSF) and the creation of new science policy (Ministry of Research, Science and Technology, MORST) and purchasing (Foundation for Research, Science and Technology, FRST) agencies. This was followed in 1991–1992 by the disestablishment of the Department of Scientific and Industrial Research (DSIR), the Ministry of Agriculture and Fisheries Technical Division (MAFTech) and the research arm of the Meteorological Service, followed by the creation of ten 'vertically integrated' Crown Research Institutes (CRIs).

The contestable PGSF was established from the funds previously allocated directly to DSIR, MAFTech and the Meteorological Service (research arm), which then had to apply for funding for their research programmes. Since 1992 the range of eligible applicants to the PGSF has widened to include the CRIs, research associations, universities and independent researchers. MORST was created to provide policy advice to the minister and to develop science and research priorities. FRST's role was to purchase PGSF science and technology outputs on behalf of government, and to provide a stream of independent policy advice.

The CRIs were created as Crown-owned companies with full commercial powers. Six of the ten CRIs had agricultural, horticultural or environmental foci, this ratio reflecting perceived needs for scientific and technological information, including those of local authorities faced with meeting substantial new responsibilities for environmental administration. CRIs have 10% of the PGSF allocated to them non-contestably as non-specific output funding (NSOF) to enable them, among other things, to manage the balance of supply and demand for research and staff expertise and to pursue novel research not sufficiently developed to submit to the main PGSF pool for funding.

An additional contestable funding scheme, the Marsden Fund, was created in 1994 to support fundamental research (not strategically directed by government), with the sole criterion for assessment being scientific excellence. This fund is projected to increase to 10% of the size of the PGSF by 2001.

An important change in the period under review in this section was the development of the government's Statements of Science Priorities. The first of these appeared in 1992 and the second in 1995 (MORST, 1995b). These priorities are used to guide allocation of the PGSF. The key direction for New Zealand science and technology is 'to foster a sustainable, technologically advanced society which innovates and adds value, especially to the strong base of biological production' (STEP Report, 1992). In the words of Minister Simon Upton: 'the focus on strategic research priorities ... makes the public science investment explicable and defensible to the taxpayers who are asked to fund it' (Upton, 1995).

The process used to develop these science priorities is of interest internationally. An expert panel guides the process with close interaction with stakeholders (through use of a delphi technique) and then utilizes an approach similar to that adopted by CSIRO to assess priorities. Such a broad-based approach to setting priorities in science and technology at the national level is not attempted in Australia. Its adoption elsewhere may provide enhanced guidance for investment in science and technology.

In summary, the changes that have occurred in New Zealand since 1975, and their implications, are as follows:

- The hierarchical model of organization was replaced by a divided model with clear separation between policy advice, science purchase and science provision.
- The public funds available for science were identified separately and made contestable to science providers.
- The rules for access to public funds were tightened and made clearer (in terms of national interest) to science providers.

- The re-grouping of science providers in subject institutes removed possible duplication and provided closer focus on results.
- Some layers of administration personnel were removed from the system, though not from the project selection process, which has become more cumbersome.
- A greater focus on non-government funding was intended by the system of purchase.
- Agricultural extension was separated off temporarily into another Crown-owned body and later sold to private interests.
- Research institutes were given tighter financial structures and financial goals were specified for them.
- Priorities were established at the national level by wide consultation and the task of implementation given to the science purchasing agency, FRST.

Australia

The period since 1975 has seen unprecedented questioning of all aspects of the research system in Australia, consequent on the country's failure to increase exports of manufactured goods to compensate for the diminished relative value of exports of agricultural produce and minerals. Also, the fragmentation of Australia's agricultural research has become quite obvious. In 1989, 493 research centres were identifiable, with an average senior staff number of 11 (Foley *et al.*, 1992). So far, change has occurred mainly by reorganizing existing institutions, and one more institution (the Cooperative Research Centre Scheme) has been created. (Another important institutional innovation in the period since 1975 has been the creation of the Australian Centre for International Agricultural Research in 1982.) This incremental approach may reflect the political constraints of Australia's federal constitution more than a difference of economic philosophy with New Zealand, for example regarding the relevance of public choice theory.

Despite the earlier triumphs of *CSIRO*, by the early 1980s it was under strong criticism, including claims that its scientists were setting their own agenda and that the organization was like an ocean liner that had run aground and no one was sure that it was worth the cost of refloating and pointing in the right direction. In addition to two major external reviews, there were numerous internally commissioned reviews of the organization's divisions and of other components of its research. At one stage, there was even a review of divisional reviews! Another review concluded that 'CSIRO in general and the rural research area in particular is over-reviewed and under-managed' (Foley *et al.*, 1992, p. xii). There have not been so many reviews since. Nevertheless, the pressure on CSIRO has continued, with a Senate Standing Committee review of its agriculture research and further internal restructuring in

1996. Altogether, there have been approximately ten major reviews, restructurings or changes of direction since the 1970s.

Some years later than CSIRO, *state departments* came under the same pressures. The Victoria department, for instance, was reviewed internally and restructured in 1985 and 1986, then subjected in 1990 to its first external review for more than half a century. In 1993 the research of the Queensland department underwent its first comprehensive external review in about a century. In Western Australia, a portfolio review, carried out in 1994 by a ministerial team with a number of members from outside the department, led to a major restructuring. Two of the main reasons quoted for change in Western Australia were a view that research was being driven by providers and that there was over-investment in production research (G.W. Robertson, personal communication). In South Australia, a commissioned review was followed by the separation of the department's research activities and relocation of some to the Waite campus of the University of Adelaide. A similar co-location of departmental research with the university has taken place in Tasmania. These two smaller states have thus moved closer to the land grant college model, but all the state departments have recently been exploring new relationships with universities.

During the 1980s, *RIRFs* were restructured, partly to make them more effective in overcoming the problems created by the fragmentation of Australian agricultural research, and partly because of complaints that scientists had captured the processes of priority-setting and allocation of these funds and that this was inhibiting necessary reform (Lovett, 1996). This experience parallels the 'capturing' by scientists of research councils set up by a number of industrialized countries after World War II. In 1985 and again in 1989 the Commonwealth Government amended its RIRF legislation. The present 'corporation' model is held to have achieved four significant changes:

1. Through its mechanism of governance, it has rebalanced the interests of end-users and providers of research in favour of users; one of the consequences has been to take a 'business system' view of research priorities and to give a higher priority to deploying new technologies for commercial benefit.

2. Through contestable mechanisms of fund allocation and creative management of the balance between competition and cooperation among research providers, it has worked to overcome the effects of fragmentation of Australian rural research capacity.

3. Through more stringent requirements of research proposals and the employment of skilled programme managers, it has levelled-up the standards of research planning, conduct, monitoring and evaluation among providers.

4. Through contestable mechanisms of fund allocation, it has provided a politically acceptable means of reallocating resources among powerful research organizations. The beneficiaries have included the Australian universities and the private sector.

Despite the perceived improvement of the RIRFs' performance, they have been reviewed four times in 5 years (Lovett, 1996).

The *Cooperative Research Centres* (CRC) scheme, launched in 1990, is the other contestable funding mechanism that the Australian Government has introduced to help overcome the structural inefficiencies and imbalances of the national research system. Briefly, the objectives of CRCs are to:

- contribute to the establishment of internationally competitive industry sectors;
- stimulate a broader education and training experience, particularly in graduate programmes;
- capture the benefits of research by the active involvement of the users of research in the work and management of centres;
- promote cooperation in research by building centres of research concentration and research networks.

Commonwealth funding of about Au$2 million (US$1.4 million) per annum is provided to each centre for up to 7 years, in the first instance. All centres must include a university partner. Agriculture and rural-based manufacturing have been prominent, with 15 of the first 62 CRCs. Another seven of those centres deal with natural resources research. The ratio of other funds 'levered' by the Commonwealth funding of CRCs is similar to that by the RIRFs. The CRCs have been as heavily reviewed as CSIRO. Each centre has been examined after 3 and 5 years, with each of these reviews in two parts. The scheme as a whole has been reviewed, even though the first centres commenced only in July 1991.

Through all the recent changes in Australia there has been a continual questioning of the rationale of governments continuing to invest in R&D to provide benefits for specific rural industries. A good deal of attention has been given to the practical definition of public goods and of the conditions under which market failure occurs. The statutory RIRFs have removed one of the chief causes of market failure in agricultural research and it is likely that Australian rural industries will have to pay a larger share of their research costs in future.

Efficiency of agricultural production

The volume of Australia's farm production has grown substantially since World War II (Fig. 7.2) despite a continued decline in the

Fig. 7.2. Indices of volume of Australian farm production (1989–1990 = 100) and farmers' terms of trade (ratio of index of prices received by farmers to index of prices paid by farmers; 1987–1988 = 100). Source: ABARE (1996).

farmers' terms of trade and the dismantling of protective measures such as subsidies and barriers to import competition. A large part of the ability of Australian farmers to remain competitive under these conditions is attributable to their adoption of new technology. One of the side-effects of reduced assistance to Australian farmers has been to focus their leaders' attention more strongly on technological innovation. Previously, much of their time was devoted to preserving government protection and subsidies.

The removal of substantial subsidies to the agricultural export sector in New Zealand since 1984 has had a dramatic structural impact, with a major decline in full-time employment in the production sector. Nevertheless, volume of production (Fig. 7.3) and value of exports have increased and processing has grown in importance as products have become more market oriented. Again, applications of science and technology are believed to have played a vital part in the ability of New Zealand agriculture to adjust successfully to the recent rapid and substantial economic changes, as well as to the earlier

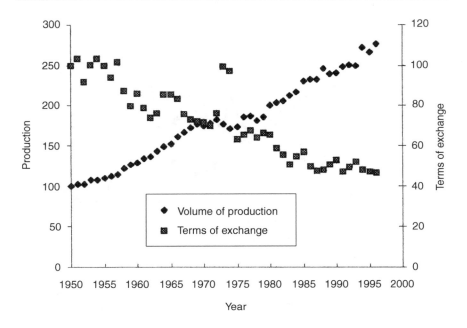

Fig. 7.3. Indices of volume of New Zealand farm production and farmers' terms of exchange (1949–1950 = 100). Sources: Hussey and Philpott (1969) and Johnson (1976, 1996).

cost/price pressures (Fig. 7.3). The impact on ANZ agriculture of the changes in research investment of the past 10 years would not be expected to show up fully until next century. Research by Scobie and Eveleens (1986) showed that, on average, the impact of research results in New Zealand peaked after 11 years and tailed off after 23 years. The time delays in Australia, with its greater reliance on crops, are likely to be briefer than those in New Zealand and probably closer to those in the USA.

Lessons Learned

The chief lesson from the NZ experience is the rate and extent of change achievable through fundamental restructuring and the development of mechanisms for contestable allocation of government research funds. Those innovations appear to have provided efficacious ways of managing change among scientists and end-users, of achieving a greater end-user focus and of ensuring that the role of science really is to underpin international competitiveness. In the New Zealand experience, the separation of policy advice from purchasing and delivery was necessary for rebuilding political trust that expenditure on science was being effectively applied to the public good (Pearce, 1995).

Even without such significant restructuring in Australia, the contestable funding mechanisms of the RIRFs and CRCs, and their exertion of considerable leverage, have provided effective and politically acceptable means of rebalancing agricultural research portfolios between commodities, regions and production and processing sectors, and among powerful research institutions. There is now also a much sharper separation between the functions of funding and providing research. It is too soon to judge whether or not some of the changes may have gone too far, for example, in reducing the proportion of uncommitted core research funds. Funding research institutions directly has the advantages of lower transaction costs, the wider casting of the net for a human activity (original research) in which individual creativity is paramount and greater attractiveness of research careers, especially for those in positions of leadership. One of the real tests of the new investment strategies will be whether or not ANZ agricultural research will be able to hold its share of the best young scientists in future.

Australians have devoted a great deal of time and effort since 1975 to the process of reviewing their agricultural research institutions. Although that process has resulted in a great deal of internal change and renaming of the units, it has not redirected or restructured agricultural research significantly at the national level. Thus, in the case of CSIRO, the RIRFs and CRCs and the setting by government in 1987–1988 of the target of 30% external funding of CSIRO within 3 years, appear to have had a greater influence on the directions of that organization's agricultural research than the two external reviews. External reviews of R&D do have other purposes, however, particularly in reassuring governments that their investments are in good hands. Nevertheless, it is arguable that many of the improvements in the efficiency of Australian agricultural R&D institutions that have occurred since 1975 could have been achieved by stronger leadership without so many reviews.

Deficiencies of the Australian and New Zealand systems

What are the deficiencies in the new investment strategies that have been introduced in ANZ? The overuse of reviewing and internal reorganizing as a mechanism of change in Australia has already been mentioned. Another weakness has been the incomplete coverage by the new investment strategies. Extension, using the term broadly for all forms of technology transfer, has been largely excluded from the new investment mechanisms in ANZ and lacks an intellectual focus in both countries, though there are some innovative groups within the universities. In New Zealand it has been recognized that extension cannot be separated from the issue of enhancing the rate of technology transfer to end-users. The New Zealand government took the view that extension services were not public goods and therefore government support was not required. The divided responsibility for technology transfer between science providers and private extension agents has left a gap compared to the former system. In Australia, state governments still fund extension, though generally at a reduced level. Some operational support is provided by the RIRFs and CRCs.

Another weakness has been in policy research. The agricultural research systems of both countries have a strong bias towards the biological and physical sciences. Some trace this back to the strong historical influences of the DSIR and CSIRO, neither of which included socioeconomic research except in minor ways. While there are academic strengths in the universities, and the Australian Bureau of Agricultural and Resource Economics provides high-level policy advice to the Australian Government, policy research is not as strongly represented in current ANZ institutions as its importance for future industry competitiveness would indicate. The lesson in this for other parts of the world is to seek broader organizational structures, as in the US land grant system, for any new system of agricultural research investment.

A third weakness has been in research on environmental issues. Though the Australian Land and Water Resources R&D Corporation, the two Crown Research Institutes (Landcare Research New Zealand Ltd and National Institute for Water and Atmospheric Research Ltd) and several Australian CRCs have major investments in research on sustainable agriculture, it has proved remarkably difficult to establish off-farm community constituencies (industry user equivalents) for such research. This point raises an issue that is likely to be important if the principles of the ANZ experience are to be applied in developing countries. The involvement of research users in the priority setting and resource allocation process is likely to be more challenging for subsistence producers in developing countries, and perhaps more like the problem of establishing a constituency for environmental research in ANZ, than it is for the relatively well-organized commercial farmers of ANZ.

Finally, there is an aspect that the ANZ agricultural research systems have never achieved, but which would be an essential part of an ideal structure. That is the full integration of the universities into the national research and extension systems.

If we were starting again

Many things would probably be done differently in ANZ, if one were starting with a blank sheet, but two of the recent changes would certainly be retained:

1. There would be some kind of centralized priority-setting and allocation system for public funds, similar to the New Zealand PGSF process. This may be about to happen for priority setting in Australia, following comments by the National Commission of Audit during 1996 that the current system is inefficient and largely unaccountable, with no clear process for setting national priorities. One of the advantages of a centralized process is that it provides the opportunity to make sure that important needs do not 'fall through the cracks' of a pluralist system. A centralized process should also be able to address the question of balance between biophysical and social science investments.

2. The second feature worth retaining would be the separation that has been created in ANZ between the purchasing and the provision of R&D, with the attendant benefits that are made possible (but certainly not guaranteed) by this separation of functions.

Asia

8

Don Mentz[1] and Gary Alex[2]

[1]Mentz International Trading and Investment Ltd,
Unit 1, 25 Longridge Road, London SW5 9SB, UK;
[2]ESDAR, World Bank, 1818 H Street NW,
Washington, DC 20433, USA

Introduction

The diversity of the countries in Asia and of their national agricultural research systems defies a concise summary of overall experience with research investment strategies. The country and research system diversity here is likely to be much greater than for other regions. The Asian NARS range in size from the two giants (India, China) to very small (Laos, Bhutan). Countries range from relatively prosperous (South Korea, Malaysia) to quite poor (North Korea, Nepal). Economic systems range from socialist (China, Vietnam) to free market (Thailand, Malaysia). Within this diversity, the subregions present some common characteristics, though even within subregions there is great diversity. One common factor for most Asian NARS has been a focus on food security as the major priority for agricultural research systems. This focus derives from the region's high populations and the concerns of the 1960s food shortages, for which agricultural technology was considered an effective and priority response. Concern with ensuring food security led to a government dominance of research institutions, which continues today, and which leads to two generalizations on Asian research systems: (i) the Asian NARS have been relatively well supported in the past and have been quite effective in introducing technologies to increase food production; and (ii) the Asian NARS are relatively conservative and have not yet undertaken major institutional and programmatic reform.

Important Changes Since 1975

In the 1960s and 1970s the agriculture/food production scene in Asia was dominated by catastrophic food shortages in many countries, with widespread pessimism and some panic concerning the ability of Asia to feed itself, then and, more particularly, in the future. The pressure was on for research solutions that would produce quick results. The International Rice Research Institute (IRRI) was established in the Philippines, and the International Crops Research Institute for the Semi-Arid Tropics (ICRISAT) in India, as the main Asian centres for this search. Generous funding was readily available from international sources and host country attitudes were very welcoming. The Asian Development Bank (ADB) was established in the Philippines, with a high profile and priority for agricultural development in general and large-scale irrigation for rice in particular. Agriculture was a top priority with most aid donors.

The results were spectacular. By the late 1970s new varieties and other technical advances, combined with major policy reforms in India and some other countries, had transformed the food production picture in Asia. Food security concerns virtually disappeared. However, progressively through the 1980s and 1990s the output of high impact research results levelled off and a series of agronomic, fertility and pest management problems appeared or intensified. Despite this, with food security no longer perceived as an immediate problem, the status and priority of research in most agricultural budgets declined, with levels stagnating or declining as a percentage of agricultural gross domestic product. Industrialization, transport, power and social projects attracted higher priorities. By the early 1990s IRRI and ICRISAT and the other more recently established CGIAR centres in Asia were experiencing severe funding cutbacks.

During the 1980s other factors emerged that were to further restrict the funding and the 'success rate' of agricultural research. These were environment and social issues, often in combination and often being linked to the 'down side' of the Green Revolution and the heavy emphasis on large-scale irrigation. Agricultural projects became more difficult and more expensive to administer and development agencies and institutions came under increasing pressure to improve disbursement rates and reduce administrative expenditures.

In the 1980s and 1990s investment in agriculture in general in Asia appears to have first stagnated and then suffered a serious decline. There is even more evidence of stagnation and decline of expenditure on agricultural research. For example, the ADB has invested more than US$12 billion in agriculture since 1967, but only 0.8% of this has been in agricultural research. Agriculture and particularly agricultural research is currently a low priority at ADB (though support appears to be stronger at the Board level). This may well be influenced by staff shortages, lending targets and the well-known difficulties and heavy staff inputs of agricultural lending. World Bank lending for agricultural research in South and East Asia has varied from year to year with no noticeable trends, and totalled US$1.1 billion in 147 projects between 1981 and 1996 (Byerlee and Alex, 1998). Some hope for the future is provided by the successes and potential in agriculture of biotechnology. The potential is clearly enormous, but there are many social, legal, technical and practical problems for the developing countries.

The main changes in research investment for agriculture and natural resources in Asia in the past 20 years have been a loss of status and priority, an overall decline in volume and possibly a decline as a percentage of the general investment in agriculture and natural resources. Some of the reasons are institutional and are due to a general shortage of grant and low interest funding. An overriding reason could be the general lack of concern in Asia over the food security situation in the foreseeable future.

Research Challenge: Outlook for Food Security in Asia

FAO (1995) estimates that the total population of Asia will grow to 4.4 billion by 2025, when there will be an additional 1.5 billion people to feed. The demand for food by the developing countries of Asia has risen/will rise at an average annual rate of 3.1% during the period 1984–2000, and total food supply in the region will have to increase by 100 million tonnes of GE (grain equivalent) every 10 years, just to maintain per capita consumption at current levels (FAO, 1992). This makes no allowance for income growth, which is a significant factor in enhancing food demand. Urbanization also increases demand for cereals and livestock products. By the year 2025, over 50% of the population of Asia will live in cities (Asian Development Bank, 1992).

Food production capacity in Asia

The best available estimates, also from FAO (1992), are that, by the year 2000, Asia will be a net importer of food. This situation is supported by claims (Brown, 1994) that between 1950 and 1984 the world grain harvest grew at a record rate of 3% per year, increasing the per capita grain available by 40%. However, between 1984 and 1993 the annual growth had slowed to 1%, reducing per capita availability by 11%. Further supporting evidence comes from data on land availability. There is very little new land available for agriculture in

Asia. What is available is of generally poor quality, with high development costs. Furthermore, there will certainly be some loss of existing high quality agricultural land to urban expansion and degradation, which is continuing at an alarming rate in some parts of Pakistan, Nepal, Bangladesh and Thailand (water and wind erosion, salinity and flooding). Significant reserves of new land now exist only in Indonesia, Cambodia, Malaysia, Myanmar and Laos.

There is also some evidence and some concern that a technology gap in food production already exists. In some areas in South Asia rice yields on research stations are making only slow progress and yields are stagnating in some of the major rice-producing areas, such as Central Luzon in the Philippines and the Punjab in India.

On this and other evidence it is now widely believed that the food production gains arising from the Green Revolution have levelled off. Some people hold hopes for significant percentage gains in the foreseeable future from the production of new rice varieties to replace the existing varieties and for other biotechnology-related advances. However, any special technology advances with widespread impact are unlikely to capture the momentum of the late 1970s and early 1980s. This scenario is accepted by FAO (1995), which views it as a major threat in Asia.

FAO also believes that, in Asia, the 1990s have witnessed some reduction in real levels of investment in agriculture. Development assistance to agriculture has certainly fallen substantially. Globally, World Bank lending to agriculture, as a share of total lending, has fallen from 30% in the 1970s to 16% in the 1990s, although to some extent this results from the declining share of agriculture in the economies of the borrowing countries. In recent years there has been a serious absolute decline in total external assistance to agriculture. In the first half of the 1980s annual commitments were about US$10 billion. This rose to US$14 billion per year by 1990, but has since fallen back to probably not more than US$8 billion (FAO, 1995, Tables 4–6). In Asia the figures were US$4.7 billion in 1980 and US$1.9 billion in 1993.

The story for agricultural research funding in Asia is probably similar, or the declines may be even larger. In terms of research for food production, the decline may be even more serious, due to the new and growing demands for research on natural resources and the environment (Pardey *et al.*, 1998b). FAO does not predict a need for massive increases in investment/research in agriculture to maintain current levels of food security; some modest increase in terms of percentage of the agricultural share of GNP would suffice, provided there are also major changes in the nature of the investment. In Asia as a whole the required annual increase in net investments is expected to decline. However, in the 1990s, investment in agriculture (and particularly that for large-scale irrigation of rice) and in agricultural research has suffered an absolute decline in Asia.

The scenario outlined above is quite persuasive. The re-emergence of serious food security problems can only be avoided in Asia by investment and technology-based intensification of food production from existing land. Current levels of investment and research funding are inadequate to avoid a pending crisis, and a further fall in current levels cannot be ruled out.

Current perceptions

Perceptions on food security issues for the immediate and foreseeable future must have an important influence on attitudes towards funding of agricultural research. In Asia, these perceptions have played a crucial role in the past and there is no reason to believe that they will not be just as important in the future. The point that needs to be made quite strongly is that the current perceptions on future food security, by policy-makers not directly involved in agriculture in governments and development financing institutions, do not appear to be in line with the facts. There is a 'fool's paradise' situation with regard to food security, with potentially serious consequences. The dangers are aggravated by the long lead times involved in taking corrective action. There is also a widespread belief among non-scientific policy-makers that research is well ahead of extension and that research can afford to mark time, i.e. there is no 'technology gap' such as FAO believes exists, but a 'technology overlap'.

Hard evidence for these views does not exist; they are merely views and the sample is very limited. However, these views could and should be tested more broadly. It may well be that an urgent need, closely following some action on the need for funding data, is to inform and sensitize those senior policy-makers in governments and development institutions, who are not directly involved in agriculture but have key roles in the allocation of resources.

Information on Asian NARS

The most recent information available from ADB is contained in a Staff Working Paper published in 1988 (Asian Development Bank, 1988). This document provides some data on research funding for various years between the early 1970s and 1987 for Indonesia, Malaysia, Philippines, Thailand, Korea, the Cook Islands, Fiji, the Solomon Islands and Tonga. No trends could be extracted, except perhaps for the Philippines where, for the years 1971 to 1983, the Philippine

Government allocation to agricultural research, as a percentage of GNP, fell from 0.083 to 0.047. A search of ADB Board Documents available to the public revealed that the Board has considered three policy papers (Asian Development Bank, 1988, 1990, 1995) related to agricultural and natural resources research and/or rural development.

There are some individual country data with various researchers. An excellent and important example of this is China where work by Fan and Pardey (1995) has been circulated in preliminary form in an EPTD Discussion Paper, and which has been used extensively later in this chapter. The same data are referred to in other documents (Fan and Pardey, 1992; Pardey and Alston, 1992). This work shows that the funding of agricultural research in China, as a percentage of agricultural GNP, has stagnated since the early 1980s. This is most important because agricultural research in Asia is dominated by China and India, with China having four times more researchers than India.

There is a serious and immediate problem with regard to statistical data on agricultural research in Asia, particularly in the area of funding. For example, specific enquiries have revealed that for two of the most important agricultural countries in Southeast Asia – the Philippines and Indonesia – no systematic compilation of research funding is available and it is not possible to draw any conclusions about trends in this regard. Many of the experienced professionals in Asian agricultural development support the view that the presently available information database for R&D is extremely inadequate to undertake the required planning for most of the countries. Hopefully this matter will be given high priority by the World Bank, the ADB, FAO, and the relevant CGIAR centres (ISNAR and the International Food Policy Research Institute (IFPRI)).

IFPRI has done some very useful work in China, which has contributed greatly to the statistical background on agricultural research funding in that country. It is also planning a major study for a much wider coverage of Asian countries. There is an urgent need to compile Asian country-specific and regional funding data for past years and to establish mechanisms and resources to maintain and update these data on an annual basis.

Asian NARS

The following section provides brief summaries of research systems in selected countries. Institutional arrangements vary widely, though there are some similarities within subregions. The authors acknowledge input and comments on this section from Derek Byerlee and drew heavily from Dar (1995), Jain (1995) and Fan and Pardey (1992).

Table 8.1. Comparison of Asian NARS.*

Country	Agricultural researchers	% PhD	% MS
China	61,000	0.27	5.7
India	20,000	—	—
Bangladesh	1,650	11.4	63.5
Indonesia	2,360	8.2	21.2
Malaysia	1,162	20.2	42.3
Pakistan	4,725	—	—
The Philippines	3,046	—	—
Sri Lanka	391	—	—
Thailand	2,012	5.0	29.0
Kazakhstan	3,000	14.3	—

* Note: From Dar (1995) and other sources mentioned in text; data from various recent years.

China

China has the largest agricultural research system of any developing country, with close to 100,000 staff (61,000 scientists and engineers) (Table 8.1) working in 1013 research institutes (Fan and Pardey, 1992; Pardey *et al.,* 1998b). The research system is almost entirely state-run. With a population of approximately 1.2 billion, of which 60% are employed in agriculture, and with limited arable land (7% of the world's arable land compared to 23% of the world population), food security has been a continuing concern for the sector and production technology has been the focus of agricultural research. The Chinese research system has contributed substantially to the agricultural sector's good performance. China has developed some first class research programmes, including the rice research programme which led development of hybrid rice technology. Chinese agriculture has reached high levels of productivity based largely on land-saving and labour-using technologies. Agricultural output grew at 5% per year from 1965 to 1990, with a jump in growth rate to 8.1% between 1980 and 1985 as a response to market-oriented economic reforms. Fan and Pardey (1992) estimate that research accounted for 20% of the growth in output.

The Chinese agricultural research system traces back to the establishment of the Baoding experimental station in Hebei Province in 1902. The system expanded as a decentralized, cooperative system of national and local agencies, which by 1949 employed a total of 470 technical personnel in a total staff of 1600. Following the revolution and formation of the People's Republic of China in 1949, the government gave serious attention to science and technology issues and by 1952, the number of technical staff in the agricultural research system had increased to about 1000 with comprehensive experiment stations in 18 provinces. Over the following 30 years, the research system was subject to a series of often traumatic reforms and influences of political movements. In 1955, a Coordinating Committee for Agricultural Research was established, followed by the Chinese Academy of Agricultural Sciences (CAAS) in 1957. Research was organized in seven regional Agricultural Research Institutes, which were placed under the CAAS, until 1958, when they were turned over to the respective provinces in which they were located. This decentralization was motivated by a goal of developing closer linkages between research and production units.

In 1958 and 1959, during the Great Leap Forward and the Anti-Rightist Campaign many research activities (including two-thirds of the staff and one-third of the institutes) were transferred to rural areas. Leaders of research programmes during this period tended to introduce unrealistic and arbitrary research objectives and work schedules. In 1962, many of these excesses were corrected, until the Cultural Revolution (1966–1976) brought further disruption. During the Cultural Revolution, many staff and facilities were again transferred to rural areas, with many scientists reassigned from laboratories to work as 'barefoot researchers' spending much time in rural labour. Total staff of CAAS decreased from 7500 to 620 in 1970. An 'open-door' research policy emphasized work on practical, field problems with involvement of farmers, and many new staff with little or no training joined the research institutes. Little systematic experimental work was done during this period and the educational system also stopped normal training programmes for scientists and technicians. In 1978, the CAAS was re-established.

The current Chinese agricultural research system is dominated by public research in national and provincial academies, prefectural institutes of agricultural science and 64 agricultural universities. The system was characterized by Fan and Pardey (1992) as a multi-ministry model. At the national level the State Science and Technology Commission (SSTC) coordinates civilian science, including allocation of research budgets. A special agency, the Chinese Academy of Science, is responsible for basic research, including biological and soil science research. The CAAS under the Ministry of Agriculture operates 37 national commodity, resources and disciplinary research institutes and coordinates basic and applied agricultural research. It has over 10,500 staff, including 5000 technical personnel. Also at the national level are the Chinese Academy of Forestry (5051 staff), the Chinese Academy of Fishery Science (3400 staff), the South China Academy of Tropical Crops, the Agricultural Environment Protection Institute and others.

Provincial academies of science, with 359 research institutions, address provincial agricultural problems and link to national efforts through collaborative programmes. At the prefectural level, 671 agricultural research institutes emphasize applied and adaptive research. Within the prefectures, an adaptive research network has over 100,000 facilities (Dar, 1995). Research coordination throughout the Chinese system is weak. A Science and Technology Department within the Ministry of Agriculture coordinates work at the national level through a rolling 5-year plan, but this has little impact on work in other ministries or at the subnational level.

The State Planning Commission at the national level is responsible for budgets and funding for ministries and the SSTC, which in turn allocates funds to ministries and research agencies. Priority setting and financial allocations for research at both the national and subnational levels lack transparency and are influenced by precedent and other non-objective factors. Funding is divided into core and project funding, with core funds used mainly for salaries. Project funds are allocated according to 5-year plan priorities with a portion allocated to proposals for work on emerging issues. In recent years, the government has reduced levels of core funds and encouraged institutes to seek funds from other sources. Production entities allocate some limited funding for research.

Human resource management in the Chinese research system presents some significant problems. Partially as a result of the Cultural Revolution, many staff joined research institutes with limited or no training. The percentage of PhDs in the system is low compared with other countries, as is the percentage of MSs (see Table 8.1). Many of the poorly trained staff have relevant skills and may provide close links between research and production agriculture, but their lack of training limits the scope for their work. Other human resource problems include the following:

- Centralized personnel and management limits flexibility in placement and transfer of staff and hinders the ability to respond to new research needs and initiatives.
- Many senior managers are due to retire in the 1990s and would normally be replaced by the next generation of staff trained during the Cultural Revolution. Since training programmes were disrupted during that period, there are limited staff with the qualifications necessary to fill these management positions.
- Large numbers of Chinese students are in training abroad, but the low salaries and difficult working conditions provide little incentive for them to return to the Chinese research system.

Agricultural universities focus mainly on teaching, though involvement in research has increased in recent years. Seven key universities administered by the Ministry of Agriculture have relatively more involvement with research, focused mainly on applied research. Provincial universities emphasize applied research and extension. In 1986, 22% of university scientific staff were identified as being full time-equivalent researchers. Under the former planned economy, though the purpose of research was clearly to serve production units and though there was a large extension system at the county level, the linkages between research and extension were weak at the provincial and prefectural levels. With market reforms, the work of extension has become more difficult in serving 170 million family farms as opposed to 6.6 million organized production units. Township (commune) extension stations are being replaced by semi-commercial agricultural technical services companies to deliver extension services to farmers.

Chinese agricultural research expenditure is about 0.4% of AgGDP, whereas the government goal for research expenditure in general is 1.5% of GDP by early in the next decade. Agricultural research expenditures are growing slowly and research intensity is declining rather than increasing. Government funding is expected to remain tight, leaving two options for increasing research funding: expanding government research institute earnings from commercial activities or expanding the current minimal levels of private sector research funding. Reforms in 1985 allowed – and encouraged – Chinese research institutes to earn income from commercial activities (in 1986, 19.1%; in 1994, 38.9%) (Pray, 1997). Following this, government funding for research declined over the period 1986 to 1995. Research institute earnings were derived from the sale of technology services, non-agricultural activities and leasing of land and facilities. Unfortunately earnings were offset by reduction in government funding and, after considering cost of commercial operations, a 100 yuan increase in earned income meant a net 81 yuan reduction in funding available for research.

Although the demand and potential benefit from technology import is high, significant barriers restrict technology transfer and private research (Pray, 1997). State-owned firms produce and market most production inputs and are protected by government. Restrictive seed laws being considered would: (i) require 51% local ownership of any seed company; (ii) restrict research to joint ventures; (iii) require a minimum US$3 million capital investment for joint ventures; and (iv) require national-level approval for joint ventures. Patent laws and plant breeders' rights laws exist, but enforcement is weak. Without significant reform of policy on private research and IPR, private sector research will be limited.

The main lessons to be learned from the Chinese experience are that:

- research results (due to an extent to the huge size of the system) can be substantial, despite severe constraints on a system;

- political disruptions or other changes affecting the human resource base for a research system have long-term implications;
- coordination is essential – but difficult – if a large system is to be efficient; and
- commercial operations by research institutions do not necessarily increase funding available to research and can in fact reduce research funding.

India

India has the second largest NARS among developing countries with over 20,000 scientists, with a solid tradition of support to agricultural technology programmes. Indian agricultural research dates from the 1800s, and received special attention following the Famine Commission of 1880. Another major boost came in 1905 with the establishment of the Indian Agricultural Research Institute (IARI) with a donation from the American philanthropist, Henry Phipps. Agricultural universities developed separately. In 1929, an autonomous Indian Council on Agricultural Research (ICAR) was established to promote and coordinate agricultural research. ICAR was reorganized in 1965, when it assumed the basis of its present structure. ICAR has now grown to include 45 central research institutes, four multidisciplinary national institutes, ten projects, four bureaux, 30 national research centres, 165 regional research stations and 84 All-India Coordinated Research Programmes (AICRP). The total number of centres under ICAR is 1300: 900 in agricultural universities, 200 in ICAR institutes and 200 elsewhere (Mruthyunjaya and Ranjitha, 1997). Much of India's research is conducted by the state agricultural universities (SAU), first formed in the 1960s and now numbering 28 with around 22,000 sanctioned scientist positions (NATP, 1995).

AICRPs are an important component of the Indian research system and predated the IARC programmes that gave a boost to rice and wheat technologies to produce the Green Revolution. The AICRPs began in the 1950s, with early success from the All-India Coordinated Maize Improvement Programme providing a model for future programmes. The AICRPs are based on a programme coordinator supported by a small number of scientists linked with a network of collaborating researchers in institutions throughout the country. The successes of the programmes are due to the central coordination, the strong decentralized network for testing, good exchange of information facilitated by annual or biennial workshops to review programmes and the cost-sharing between ICAR and the collaborating state institutions. (ICAR funds 75% of costs with university collaborators and 50% with other state institutions.)

Indian agricultural research funds are provided by the central government (60%), state governments (20%), the private sector (12%) and others (8%). Despite the fact that 12% of agricultural research is funded by the private sector, in the past linkages between the public and private sector were weak, and official attitudes to the private sector occasionally perceived as hostile. This situation is changing rapidly.

Pakistan

Pakistan's agricultural research system is based on 74 research establishments at the federal level, 106 at the provincial level and three agricultural universities engaged in research. Each province has its own research capacity with overall coordination from the Pakistan Agricultural Research Council (PARC). A 1988 personnel survey identified 4725 scientists and 1739 support staff in the system. Approximately 35% of the scientists were at the federal level (PARC National Master Agricultural Research Plan). PARC was created following a 1978 review of the Pakistan research system to promote and coordinate an integrated federal–state research system. PARC itself has substantial research implementation capacity, whereas provinces have independent research infrastructure funded largely with their own resources. Research organizational structure varies among the provinces with the North Western Frontiers Province having integrated all research and extension under the agricultural university.

Funding is a serious problem, as research institutions rely almost entirely on public funds and research institutions are seriously underfunded. Currently, the ratio of salary to operating costs varies around an unacceptable level of 85:15, a situation that puts additional stress on organizational structures for research (Nickel, 1997). Plans for a competitive research grants programme may help to ameliorate this situation.

Bangladesh

Bangladesh has 1650 scientists, the third largest agricultural research system in the subregion. At independence in 1971, Bangladesh already had a long history of research for traditional export crops (tea, jute), but relatively little research infrastructure for other crops. The country moved quickly to establish ten research institutes, including the largest, the Bangladesh Agricultural Research Institute (BARI), with more than 600 scientists and the Bangladesh Rice Research Institute (BRRI), which is one of the world's largest centres for rice research.

Eight of these institutes were set up as autonomous institutions with a Director-General reporting to the relevant government ministry. Agricultural universities, primarily the Bangladesh Agricultural University and the Postgraduate Institute of Agricultural Education, also conduct research.

The Bangladesh Agricultural Research Council (BARC) was established as an apex organization to coordinate research, strengthen local research capacity, promote research investment and assist with strategic planning. BARC has no direct funding or management role or control over research institutes, which may report to different parent ministries. This arrangement has made it difficult for BARC to coordinate research activities effectively, though its influence has increased when it has had access to donor project support for research.

Nepal

Nepal established its National Agricultural Research Council (NARC) in the late 1980s to coordinate research in a complex network of regional stations, commodity programmes and disciplinary divisions. The NARC coordination and the research system as a whole have not been fully effective and a restructuring of the system is now taking place (Jain, 1995). The NARC is being reorganized as a more autonomous institution to define research policies and priorities, allocate resources, monitor research results and coordinate research programmes. NARC will be controlled by an Executive Board, chaired by the Minister of Agriculture and including representatives from three government departments, the university, agribusiness, farmers and scientists.

The Council is reorganizing research facilities into two new institutes: a National Agricultural Research Institute and a National Animal Sciences Institute. In addition, research stations are being reorganized to form four regional stations with 16 additional stations. The NARC will serve as the apex research organization for the country. The agricultural university institutes – the Institute of Forestry and the Institute of Agriculture and Animal Science – have considerable scientific human resources and conduct some research, but remain outside of the national research system coordinated by NARC.

Sri Lanka

Sri Lanka has a large number of separate research institutions (crops, export crops, rubber, tea, coconut, aquatic resources and others) with the largest of these being the Department of Agriculture under the

Ministry of Agriculture. The Department of Agriculture with about 250 scientists has a network of regional research stations and is organized into institutes for rice, horticulture and field crops. Some of the strongest research institutes are those for single-commodity export crops (tea, rubber and coconut). These have a long history of support to commercial, export agriculture. They were initially and are again becoming increasingly private sector in nature. These multidisciplinary institutes benefit from a clear commodity focus, close linkages to clients, a degree of operational autonomy, and relatively stable support from a local tax on exports or other industry support (Jain, 1995).

The Faculty of Agriculture of the University of Peredeniya and the Post-graduate Institute of Agriculture, as well as some smaller agricultural faculties, have a substantial, well-qualified staff which undertakes some research. The large number of research institutions with a relatively small number of scientists results in some inefficient operations, duplication of programmes and facilities and difficulties in coordinating research. For this reason, the Council for Agricultural Research Policy (CARP) was established in the late 1980s to coordinate research programmes and funding and improve system efficiencies. Though the Council has introduced some management improvements at the margin, it has not had major impact on the system as it has not had budget authority over research institutions and the research establishment has proven to be quite conservative.

Lessons from the South Asia experience

- Top level councils may be a useful mechanism for promoting, coordinating and monitoring research, but such councils can be structured in various ways and clearly do not solve all problems with research management.
- Proliferation of research agencies and stations, either in a federal–state system or in a unified national system, leads to inefficiencies and difficulty in coordination.
- Universities can play an important role in research systems but are too frequently left out of the coordinated national system.
- Incentives are needed to encourage private sector research and greater public–private partnerships.
- Research management requires mechanisms to shift research focus proactively to new priorities and strategies for productivity increases.

Southeast Asian NARS

Malaysia

Malaysia has introduced interesting reforms and positioned its agricultural research within overall government development strategy (Dar, 1995). In Malaysia, the agricultural sector has been important, contributing 23% of GDP and 37% of employment in 1990 (Hashim, 1992). Scientific research to support the agricultural sector began in the early 1900s with the beginning of rubber research. Following independence in 1957, the country moved to diversify research capacity to support production of food and other export crops (see Table 8.2). Institutions are charged with both research and development and in the case of commodity institutes have close linkages with industry.

Table 8.2. Institutional make-up of Malaysian NARS.

	No. of scientists	Support staff
Rubber Research Institute of Malaysia	196	1857
Malaysian Agricultural Research and Development Institute	474	3120
Forestry Research Institute	91	354
Palm Oil Research Institute	108	449
Malaysian Cocoa Board	—	—
Veterinary Research Institute	30	170
Fisheries Research Institute	75	—
Universiti Pertanian Malaysia	250	—

Source: Hashim (1992).

Malaysian government policy, outlined in the 'Vision 2020' strategy, envisions Malaysia attaining industrial country status by the year 2020. Government strategy projected that AgGDP would triple by 2010, but would then contribute only 10% of total GDP, while agricultural employment would decline only slightly. The development agenda calls for agriculture to provide food self-sufficiency and contribute to balanced growth. Significantly, the 2020 Vision identifies science and technology as one of nine challenges for the country to achieve the vision for 2020.

Malaysia's research policy is guided by the National Council for Scientific Research and Development (NCSRD), established in 1975 to advise the government on science and technology issues and to coordinate policies and support to this sector. NCSRD operates through two committees: a Standing Committee on Science and Technology Development and Management and a Coordinating Committee on the Intensification of Research in Priority Areas (IRPA). Agriculture is one of the research priority areas, and in the Sixth Malaysia Plan (1991–1995) agriculture received US$273.8 million or 45.6% of funding for direct R&D. The Ministry of Science, Technology and Environment serves as secretariat to the NCSRD and implements approved policy and strategies.

The IRPA mechanism established in 1987 provided more systematic coordination and accountability for research programmes. National priorities for agricultural research are determined by the IRPA Coordinating Committee. Research institutions then formulate projects within the priority areas, subject these to in-house reviews and forward them for NCSRD review and approval, with approvals on a 5-year basis. Research programmes are monitored both in-house by the research institute and by IRPA panels, which review projects twice a year. Malaysian science policy gives priority to technology transfer and commercialization. Policy calls for research institutions to self-finance 30% of their operating budgets by 1995 and 60% by 2000. A technology park and a technology corporation were established to facilitate technology commercialization. The recent reforms of the Malaysian agricultural research system emphasize:

- linkage between research and development to ensure the impact of science on government objectives;
- relating research and science policy to national objectives;
- monitoring and evaluation of research programmes;
- international and local research collaboration;
- staff development, performance appraisals and staff incentives; and
- encouraging self-financing through contract research consultancies and other sources of revenue.

The Philippines

Philippine agricultural research began with establishment of the Bureau for Agriculture in 1910, but expanded in various government agencies and received limited government support. In the 1950s, the state colleges and universities, principally the University of the Philippines College of Agriculture, intensified efforts in agricultural research. In 1972, a review of agricultural research characterized the research system as fragmented and with little impact on economic development. This led to the establishment of the Philippine Council for Agriculture and Resource Research and Development (PCARRD) to formulate policies and research programmes and guide their development. In 1987, government reorganization limited the authority of PCARRD (which is under the Department of Science and Technology) to coordinate research activities under the Department of Agriculture. At that time, the Bureau of Agricultural Research (BAR) was established to coordinate Department of Agriculture research activities.

Research responsibility is split, with PCARRD and the state colleges and universities generally responsible for more basic research and the BAR responsible for applied and developmental research. The state colleges' and universities' leading role is due to their strong human resource base, which contrasts with the BAR strength in its network of research facilities. Current research in the Philippines is executed by:

- the Department of Agriculture with 46 national research centres, 88 regional research facilities, 13 Regional Integrated Agricultural Research Centres and 75 regional outreach stations;
- state colleges and universities, of which 39 are engaged in agricultural research and which work on six national and six regional research centres maintained by PCARRD;
- provincial governments which operate 143 provincial facilities for technology verification; and
- private sector firms focusing on plantation crops and seed and other input technologies.

BAR provides direct coordination for Department of Agriculture research and maintains close liaison with PCARRD through the umbrella National Agriculture and Resources Research and Development Network (NARRDN). This network incorporates national and regional research centres, BAR, PCARRD, state colleges and universities, private agencies and others involved with research. For improved coordination at the regional level, PCARRD formed Regional R&D Consortia in the 13 regions of the country. The consortia include regional state colleges and universities, research stations, BAR, PCARRD and others.

Indonesia

Indonesian agricultural research had its start in the late 1800s with Dutch establishment of the Estate Botanical Gardens and single-commodity experiment stations for major export crops (tea, coffee and sugar cane). The system grew during the 1900s with most Indonesian research under government departments. In 1974, in an attempt to improve research cooperation, the government established the Agency for Agricultural Research and Development (AARD) under the Ministry of Agriculture. AARD was reorganized in 1984 and given a mandate to: plan and coordinate research within the Ministry of Agriculture; formulate policies to guide research; manage research centres; control research units under the Ministry of Agriculture; and evaluate research findings and performance. AARD consists of a Secretariat, two centres, five commodity coordinating centres, 16 research institutes, 40 research stations, 63 laboratories and 181 experimental farms. It also provides financial support and direction to six estate crop research centres.

AARD has linkages with agricultural universities and has established a competitive grants programme to channel funds to university researchers. Research strategy is closely linked to development efforts with applied research programmes bringing researchers together with local extension agents and farmers for technology verification in Regional Agricultural Technology Assessment Institutes. Indonesia has made substantial progress in expanding its research system. Emphasis on training and staff development increased the number of PhDs from 399 to 933 between 1984 and 1993. In addition, the research programme diversified its agenda between 1977 when 70% of scientists were working on food crops (mainly rice) to 1992 when over 50% of scientists were working in areas other than on food crops.

The Indonesian government provides strong support to the research system and generally protected research funding levels during a period of economic adjustment (Tabor and Alirahman, 1995). The research system has been effective in supporting increased production, including the attainment of rice self-sufficiency. Between 1968 and 1983 rice yields nearly doubled to 3.9 Mt ha^{-1} per season, and since then rice yields have continued to grow, though at a slower rate. Also between 1983 and 1992, maize yields increased by 28%, soybean by 44% and cassava by 24%. Annual yield growth of oil palm was 3.8%, tea 4.8% and cocoa 13%. The Indonesian rice integrated pest management (IPM) programme became a success story for reducing pesticide use and increasing farmer incomes (Kenmore, 1996). In supporting Indonesia's current 25-year plan, AARD has seven research policies:

1. Development of location-specific technologies to support development goals.
2. Development of specific commodities on a regional basis.
3. Diversification of production to meet new consumer and industry requirements.
4. Increased production efficiency to reduce labour requirements.
5. Encouragement of a shift to commercial agriculture.
6. Promotion of balanced development between regions.
7. Preservation of the natural resource base.

Thailand

Thailand's research system is diverse, reflecting the country's important and dynamic agricultural sector. The research system is split between government and a diverse private sector. The Department of Agriculture (DOA) under the Ministry of Agriculture and Cooperatives conducts the bulk of the traditional field crop research though there is also government research under other departments (Livestock, Irrigation, Land Development and Fisheries) and in the universities. The DOA research system has 25 regional research centres and 67 research stations managed by five crop institutes (rice, field crops, horticulture, rubber and sericulture) and other departments have an additional 57 research stations. Under the DOA, regional R&D units at eight regional research centres are charged with providing laboratory services, promoting technology transfer and co-ordinating research between agencies. At the conclusion of a World Bank project in 1994, the regional centres had 631 scientists, 30 with PhDs and 183 with MSs. In addition to this decentralized research capacity, the DOA has technical disciplinary divisions based in Bangkok.

The private sector operates quite freely and technology import is important for much of Thailand's varied commercial agriculture. Biotechnology investment by the private sector has been substantial. As the country's economy becomes more sophisticated, the agricultural sector must meet demands for higher quality and more diverse goods for the local market and provide a higher value export product range. It is questionable whether the research system can provide the strategic research needed to generate new technology for a more complex product mix and whether the country's reliance on technology importation is too great.

Lessons from the Southeast Asian experience

Southeast Asian research systems have developed with government support focusing on generation of public goods technologies for basic food crops and have generally been quite successful in providing a base for increased food production and rice self-sufficiency. Lessons that might be drawn from experience here are that:

- linking research to development provides a focus for researchers, a clear purpose for funding from government and may help to maintain a stable level of funding from government; and
- organization of NARS under apex autonomous agencies seems to provide more coherence and better coordination of research than

does organization directly under ministries or under coordinating committees.

Central Asian NARS

The countries of central Asia (Kazakhstan, Uzbekistan, Turkmenistan, Tadjikistan, Kyrgyzstan) and their national agricultural research systems are – as is the rest of the former Soviet Union – struggling with adjustment to a market economy. Reforms are needed for institutions to live within current budget realities and to meet needs of the emerging private sector. The agricultural technology systems of these countries have a high potential for success in introducing new technologies due to: their large cadre of well-trained scientists; government commitment to the agricultural sector; and their potential to draw on regional and international technology programmes (CGIAR, 1996a). At the same time, serious constraints to these research systems are: their research policies which still reflect past conditions of central planning; a need for major organizational changes to allow for more efficient and effective research; severe funding constraints; lack of effective linkages to farmers; poor morale among scientists; and outdated and deteriorated infrastructure. Food shortages in the region have been common and food security is a priority for research. Countries lack policies for diversification and technologies and research programmes to support new crops. Environmental problems are severe and will require increased attention to developing sustainable management systems for natural resources.

Kazakhstan, the largest country of Central Asia, provides an example of the central Asian NARS. Kazakhstan became independent in December 1991 and embarked on an ambitious programme to restructure its economy and build an effective government and a national identity. Since then, Kazakhstan has made substantial progress in transforming its economy into a market system and has reshaped its political system, though economic adjustment has been difficult. Between 1991 and 1995 the GDP declined by 50% and government budgetary resources declined by 75%. The country enjoyed relatively high levels of public services, but these can no longer be maintained with current economic and government fiscal conditions. It now faces wrenching demands in rationalizing government services and reducing government employment, an adjustment which will affect the scientists employed through the Academy of Science (World Bank, 1997a).

The Kazakhstan agricultural sector is characterized by its potential, its low productivity and its state of transition. The country's abundant land (291.3 Mha) and moderate population (17.5 million) allow for a wide range of crops and livestock, but production potential is limited in that more than 80% of the land is in steppe and desert, the climate is continental with extreme summer and winter temperatures, and rainfall is distributed unequally. The country has approximately 33.5 million sheep and major crops are wheat (12.8 Mha), barley (7.0 Mha), rye (0.6 Mha), millet (0.5 Mha), oilseeds (0.4 Mha), and vegetables and potatoes (0.3 Mha). Planted areas have been determined by government plans and adjustments can be expected as producers begin responding more to market forces.

The agricultural sector contributed 20.2% of net material product (NMP) in 1993, down from 30.4% in 1992, 34.2% in 1991 and an average of 36.2% from 1986 to 1990. The sector employs 22.6% of the workforce, with employment having decreased slightly in the last 3 years. Agricultural productivity is low, reflecting natural (soil and climatic) limitations, inappropriate technology use and lack of support from efficient markets for inputs and products. For six major crops (wheat, barley, corn, rice, sugarbeet, potato) Kazakh yields averaged only 40.8% of world average yields in 1996.

Transition to a market economy is the greatest challenge facing Kazakh agriculture. A move from state-planned and state-operated agriculture to a market agricultural system was to be led by an expedited programme to privatize 2485 state farms, 5000 output marketing enterprises and 700 input marketing enterprises supporting the agricultural sector. Privatization has gone ahead, with approximately 98% of state farms said to have been officially privatized. Of this, however, 90% of the total land is still operated under state control as in the past. Approximately 8% of total land is under truly private management. Marketing enterprises too are privatized, but still locked into old patterns of operation and old commitments. The agricultural sector has thus had subsidies eliminated or reduced and state support withdrawn, but has not been freed to respond to new market opportunities and adopt production efficiencies. Production has declined for many crops, as have farmer incomes. Disenchantment with economic reforms is said to be high in rural areas. Action must be taken soon to improve rural conditions and prevent more severe social problems and unrest in rural areas.

The National Academic Centre for Agrarian Research (NACAR) was established in May 1996. It was formed from the Kazakh Academy of Agricultural Science, which traced its origin back to the founding of the Kazakh branch of the Academy of Agricultural Science of the Soviet Union in 1941. The Academy has 34 scientific research institutions, 28 experiment stations, 50 experimental farms, scientific production unions and a staff of over 3000 scientists with 210 DSs and over 1000 PhDs (ICARDA, 1997). (In comparison, the US Department of Agriculture's Agricultural Research Service has 2600 scientists and all US State Agricultural Experiment Stations have

6400 full-time-equivalent scientists.) The Academy is organized into six departments: economics and information; crop production; natural resources; livestock; mechanization; and postharvest handling.

The Academy recognizes a serious problem with scientist morale, as professional prestige declines and as support for work of the Academy decreases. Senior scientists are retiring and younger scientists are leaving for better-paying occupations. Financing is a severe problem for the Academy, with salaries taking a large share of budget and very little (0.3%) allotted for equipment procurement and facility upgrading. Other central Asian research systems generally face problems similar to those of the Kazakh NACAR. The central Asian academies of agricultural science suffer from:

- excess staff and facilities above that which can be supported by expected future budgets;
- lack of interaction of scientists with producers and agribusiness;
- highly qualified staff frustrated with deteriorating terms of service and unable to work because of lack of operating cost funding;
- staff proud of past Academy traditions and prestige and resistant to change;
- inadequate focus on socio-economic research;
- lack of technology dissemination channels and strategies to reach private farmers and agribusiness; and
- need to shift greater attention and resources to environmental and natural resource management issues, especially for development of sustainable production systems.

Lessons from the Central Asian experience

The central Asian research systems face the following significant challenges:

- First, the systems will have to restructure themselves as efficient technology innovation systems to serve the needs of a dynamic private sector agriculture. This will entail privatization and divestiture of excess facilities, re-orientation of programmes to emphasize technical innovation on private farms, establishing a role for producers and agribusiness in the governance and financing of research, and establishing a technical advisory service system for private producers.
- The research systems must support reform by facilitating the transition of many former agricultural labourers into private farmers motivated by and able to respond to market signals. This may require a large-scale training effort and substantial advisory services on business planning and marketing, as well as technological

innovation. Research institutions must also respond to technology needs of emerging agribusinesses and integrate these businesses into the research system. Agricultural universities will need to revise curricula to produce graduates oriented towards private sector agriculture and grounded in business and management skills.

- Research institutions will have to adopt programme priorities to facilitate the transition of the former state farms to private farms; address environmental problems, especially salinization and erosion; increase productivity on higher potential lands; and develop new sustainable farming systems, including rotations, matching of crops to production zones and better use of water resources.

- The countries of central Asia, which share many common conditions, history and institutional structures, should exploit this potential for regional collaboration. A December 1995 workshop identified priorities for collaboration on: genetic resource conservation (recognizing the importance of the Central Asia region as one of the Vasilov centres-of-origin for many crops); varietal improvement for cereals and legumes, especially for resistance to abiotic stresses; farm resource management to address water use efficiency, salinity, water erosion, economic transition and soil fertility decline; range management and livestock marketing; and seed production and marketing (ICARDA, 1995).

Challenges for the Future

Increasing the funding of agricultural research

In these times of resource scarcity, particularly of concessional development assistance, and the growth in competing demands (including those within the agricultural sector itself), there is a need for special action to build a funding constituency for agricultural research. This is particularly the case in Asia, where the common perception is one of prosperity and high growth rates. This need is urgent. Whatever the mechanism for action, a regional focus would seem to be highly desirable and FAO, the World Bank or the Asian Development Bank should facilitate a dialogue on these issues.

The formation of the Asian Pacific Association of Agricultural Research Institutes (APAARI) is a hopeful development and APAARI is emerging as an effective regional organization. To date its main value has been as a forum for discussion for all those involved in agricultural R&D in the region, including donors. This ensures that there is an awareness in all member countries of their respective R&D agenda, and the opportunities for generating active partnerships in project work. The donors are finding this particularly valuable. Perhaps, in the future, the role of APAARI could expand to include assisting in the brokering of consensus on policy and strategic issues.

In the current circumstances, to seek more of the same of the past 20 years would not be realistic or desirable. Rightly or wrongly, a good deal of the agricultural development investment of the past has been discredited. This is particularly so for the major irrigation development projects, which dominated Asian Development Bank agricultural sector lending in the 1960s and 1970s. In the 1980s, the fact that these projects made major contributions to food security tended to be overshadowed by the growing wave of criticism of such projects for their perceived adverse social and environmental impact. These perceptions are real and would appear to be still relevant in the overall attitude to agricultural research funding by policy-makers. However, for the record, it must be said that these major agricultural development investments by the development institutions are in fact being greatly undervalued, particularly as far as Asia is concerned. Irrigation development in the region expanded from 85 Mha in 1986 to 137 Mha in 1991, an increase of 61%. As a result of this area expansion, backed up by the application of science and technology (mainly higher yielding varieties and fertilizer), the FAO food production indices for 1979–1981 of 100 rose to 129.18 in 1994. For the world as a whole they were 103.16 for 1994 (F. Tacke, 1997, personal communication).

To meet these challenges, what is needed now is some revision of objectives and strategies, a major revision of presentation and approach and a new wave of enthusiasm and commitment from all concerned. The research behind the Green Revolution was strategy based (produce the miracle plants and the investment and production will follow), enunciated by the founders of the CGIAR and vigorously pursued through the international centres. However, for Asia, at the moment, there is no broad strategy for agricultural research and only small pockets of enthusiasm and commitment, which have not succeeded in motivating those at senior policy levels in governments and financial institutions. Apart from any cooperative regional action to define a research investment strategy, and promote support, there are a number of strategies that could be taken to promote research investment. Some examples are discussed below.

The FAO view: farmer financing

FAO (1995), through the World Food Summit and other activities, has been attempting to sensitize the world to food security issues and the need to arrest the decline in investment in agriculture and natural resources. Although research gets a mention, it has not been a focus of recent FAO efforts. In global terms, the FAO view is that investment policies and strategies must be adopted that eliminate and reverse the current investment bias against rural development and which favour domestic saving and investment, especially at the rural household level. More specifically, the FAO view is that:

> Investment (in the future) in agriculture will be funded primarily from domestic savings at the farm household level. The public sector and external assistance are neither able, nor are they suitable, to replace the farmer's own contribution towards raising food output, but they have essential catalytic functions in shaping the economic and institutional farming environment through the provision of public goods and social services.
>
> (FAO, 1995)

All the practical indications suggest that this broad global concept from FAO is highly appropriate to Asia and that the implications run directly through to the agricultural research subsector. Great progress has already been made by FAO in Asia in developing the concept, evolving strategies and methodologies and guiding broad-scale implementation. The initial focus and catalyst was insect-related crop losses in rice and integrated pest management (Box 8.1), but the application and implications are much broader.

Increasing donor investment in research

Donor agencies in particular, but also Government and private investors, have a ready-made and relatively simple means of improving the returns on their investments in agriculture: by increasing the share devoted to research. This matter has been studied widely and investment in research has been shown to generate consistently higher returns than total investment in agriculture. Furthermore, the security of agricultural investments is improved by an appropriate research investment component. Echeverría (1990) has compiled the results of studies, estimating the returns from investments in agricultural research for various commodities on a global scale from 1958 to 1990. For rice, the annual internal rate of return ranged between 60 and 96%. This was confirmed by work at IRRI in the 1980s (World Bank, 1992).

Despite this evidence, institutions such as the Asian Development Bank have given surprisingly little emphasis to research in agriculture (only 0.8% of total investment in agriculture). If current percentages could be doubled, the impact on research funding would be dramatic and there would be no problems of diversion of scarce resources from

Box 8.1. Rice integrated pest management: case study.

Many now accept the need for research strategies and funding that provide much more support than in the past to domestic savings and investment at the rural household level. Some believe this to be the key to future food security in Asia. In Asia, the work of FAO on integrated pest management (IPM) in rice has already provided a large and most valuable pool of experience. National IPM programmes now exist in the Philippines, Indonesia, Vietnam, Malaysia, Sri Lanka, Thailand, Republic of Korea, People's Republic of China, Cambodia and Bangladesh.

The relevance here of the IPM rice programme is in the strategies and techniques in working with farmer groups, encouraging self teaching in technical and scientific matters and farmer decision making on technical and investment matters. The work has experienced great success: high popularity with farmers in many countries, significant policy reforms by a number of governments including the Philippines, Indonesia, India and Vietnam, and major investment decisions by governments in support of the activities. The Government of Indonesia borrowed US$40 million of development bank non-concessional funds, to fund the organization and promotion of IPM activities. Government funding is often necessary to kick-start the IPM activities at farmer level, which lead to the relatively huge farm level investment. This is a persuasive demonstration of the extent of acceptance of these farmer-based activities, at grass roots and government levels.

Thus, a major objective of any Asian strategy for agricultural research might be, in parallel with continuing support for new technologies, the need to encourage much greater on-farm saving and investment, supported by a high priority in the allocation of research investment to research which is of direct relevance and interest to farmers in their investment decisions.

other sectors. Most donor and development funding agencies have opportunities to encourage national government cooperation in such matters and the policy arguments to support the establishment of the necessary guidelines are sound.

Private sector funding

Traditionally, there have always been substantial private investments in agricultural production activities in Asia – coconuts, oil palm, rubber and more recently pineapples, bananas and other tropical fruits – and these have supported considerable research and development activity. Apart from this traditional investment, in recent years there has also been a large increase in foreign private capital flows into some Asian countries. Generally speaking, these flows have bypassed agriculture. Today, agriculture remains a relatively unattractive sector to private investment (FAO, 1995).

The situation for agricultural research in Asia is similar, but there is some evidence over the past few years that the private sector is becoming more involved in agricultural research. This still does not involve a large proportion of the funding available, and what there is appears still to be largely through grants to public sector institutions. Where there are signs of greater private sector involvement, it is mainly in areas such as biotechnology. Other than funding from foundations (which is now rather limited), there is still not much funding from the private sector for the international agricultural research centres (G. Rothschild, IRRI, personal communication). Part of the problem is that in most developing countries of Asia it is difficult for the investor in research to secure the products of the research as return on investment. Furthermore, there are many who believe that public funding of research, particularly in developing countries, permits greater attention to the public interest. Policy action by governments to improve the incentives and opportunities for the private sector to invest in potentially profitable research and development projects in the agricultural sector could have a useful impact.

In the search for innovative ways to generate private sector funding for agricultural research, the possibilities for indirect private funding, through negotiated public/private partnerships in area development schemes, may be a vehicle worth revisiting. This would be similar to the integrated rural development projects, which have not been popular in the 1990s due to poor performance, and low returns from such projects in the 1970s and 1980s. However, this poor performance is now thought to have been mainly due to operational complexity and to a top-down approach to planning and management that overtaxed the management capacities of the implementing public sector authorities. Results are expected to improve in the future, with decentralized financing and greater participation of the stakeholders in planning and implementation (FAO, 1995; World Bank, 1995b).

The concept is to involve the private sector in multifaceted programmes, typically with a number of integrated projects, which provide opportunities for investment of all types, from government infrastructure to fully commercial production, processing and marketing, NGO activities, etc. In the funding negotiations the private sector would be encouraged to accept responsibility for funding public services (such as health and education) as a *quid pro quo* for government or donor funding and government execution of research. All sorts of refinements are possible, such as substantial private sector input into the research programme, evaluation processes and so on.

This concept deserves serious consideration by the development assistance community because of the potential for assisting minority groups and disadvantaged inland and mountainous regions. Often the main problem is a shortage of economically viable development possibilities and central governments that are more interested in the more prosperous, industrialized coastal areas which have large and growing infrastructure requirements and a strong political voice. Many of the inland, mountainous areas are also economically depressed and have serious environmental problems due to shifting cultivation and expanding areas of treeless hills still designated as 'forest lands'. Many development opportunities with these characteristics can be found in Asia, particularly in China, Laos, Cambodia and Vietnam. It is difficult to devise individual solutions for the many problems. One of the problems is the need for research, but this may not materialize to the extent required without external assistance. In current circumstances external assistance will not be available without a clear economic viability objective and prospect. Hence, the vicious circle continues. In these circumstances the international private sector is needed, with international markets and the technology and volume to deliver quality at competitive prices.

There are many examples in Asia, some dating back three or four decades, of agricultural development programmes funded jointly by government and the private sector. These often included processing, research and social components. They have mainly been in intensive production circumstances (oil palm, pineapples, bananas, etc.) and in the more favoured areas of the Philippines, Malaysia, Indonesia and Papua New Guinea. Generally speaking, the difficult problems in the vast areas of Asia which are disadvantaged and depressed in economic terms – China and Indo-China in particular, and where some of the greatest environmental challenges exist – have hardly been touched. Research, and the sustainable funding of research, must be key elements in any future development strategy. If governments, the private sector and development institutions do not work together to produce integrated solutions, development in these areas will be slow to get started and will be prejudiced by small-scale, *ad hoc*, exploitative commercial ventures, and the human and environmental problems will continue to deteriorate. In this respect, those promoting the case for research funding should also lend support to a revised research agenda to encompass the broad national needs.

Networks

Many believe that, for the foreseeable future, the real levels of funding to agricultural research in Asia are likely to decline, or at best stagnate at current levels (Pardey *et al.*, 1998b). This may be a realistic view. In such circumstances, efforts to improve the effectiveness of investment in agricultural research (particularly investment by development assistance and aid organizations) would be critical. This is another angle to the research funding dilemma that might well benefit from much closer cooperative study and action by the organizations concerned. One focus that might emerge from such an approach is the high returns achieved from investment in regional and subregional networks for agricultural research, including exchange of information, scientific consultation, collaborative research and exchange of research products. Such activities have been found to be extremely valuable and productive and have proven to be a suitable vehicle for external expertise and financing, particularly in the early stages. In 1988 there were more than 30 regional or subregional networks on agricultural and related subjects in Asia, almost all of them involved in or associated with research in one way or another (Asian Development Bank, 1988). The current total would be many more, although it would appear that no comprehensive study or even listing of networks in Asia has been undertaken in recent years.

Since the late 1980s the ADB has been particularly successful with technical assistance grant funding support of a number of networks in Asia for the systematic exchange of germplasm. For South Asia the phase 1 programme was for six countries: India, Pakistan, Nepal, Bhutan, Bangladesh and Sri Lanka. This phase 1 laid the groundwork for the exchange of germplasm of the principal crops, and for assessing performance on yield and disease/pest control. Similar assistance is being given to an Indo-China network: Cambodia, Laos and Vietnam (with Swiss funding for Myanmar). The ADB has also funded two phases of a programme for Southeast Asia: the Philippines, Thailand, Indonesia and Malaysia. This group is now considering self funding for a third phase. There are few examples of direct funding from developing countries, but some direct support has recently come from India. In all networks the developing country partners are expected to contribute in kind. The ADB support in this area has been systematic and long term. It has been through the Asian IARCs: IRRI, ICRISAT, IIMI, AVRDC, IBSRAM, ICIMOD and IJO; and the Asian outreach programmes of CIMMYT, CIAT, CIP and IPGRI. The IARCs and the NARSs have both derived benefits from the networking arrangements. The IARCs have provided linkages to international research programmes.

Networking has been found to be effective in Asia in conducting research, provided the linkages established lead to a clearly defined agenda, with sound projects, a clear division of labour between partners and maximum 'spillovers' to all concerned. At IRRI, networking for entire production systems is organized through the research consortia for each rice ecosystem: irrigated, rainfed lowland, upland and flood prone. Networks have also been established for particular areas of research such as IPM and rice biotechnology. These incorporate, among other things, a strong capacity-building component. The Asian Rice Biotechnology Network (ARBN) and the Rice IPM Network (IPMNet) are good examples. The Asian Rice–Wheat Cropping System Network is another good, recent example. The most extensive network has been the International Network for the Genetic Evaluation of Rice (INGER) which involves almost 100 countries in all continents. This network has underpinned the improvement of rice varieties in all these countries.

Administrative Issues

This discussion of practical remedies would not be complete without mention of the administrative problems associated with agricultural research. From a management point of view, agricultural research is difficult, staff intensive and hard to evaluate in the short term because of the social factors and other complexities. In current circumstances, in most aid and development institutions, management staff are being pressed strongly on administrative economies, usually through broad measures such as staff ceilings and administrative budget restrictions. Therefore complex agricultural development/research projects and difficult policy areas such as agricultural research strategies tend to be neglected. This problem may well need to be addressed by positive discrimination, in terms of management focus and administrative resources, in favour of agricultural and natural resources development and research.

Conclusion

In the Philippines there was a doubling of government funding for agriculture (including research) in 1996–1997, triggered by shortfalls in domestic rice production. Hopefully, the efforts to generate more funding and identify new sources for funding agricultural research worldwide, will be successful well ahead of the re-emergence of serious food security problems. Given the normal lead times necessary for agricultural research outputs, notwithstanding the advances in biotechnology, this lead time will be extremely important to meet food production challenges. For Asia, an early start in dealing with the pressing need for more data to support planning and decision making would seem to be a sensible and relatively inexpensive first step.

Research Investment Strategies in Africa: Lessons Learned

Sub-Saharan Africa

9

Mohamood A. Noor

SPAAR Secretariat, World Bank, 1818 H Street NW, Washington, DC 20433, USA

Historical Background

Colonial period

In the late 19th century, agricultural research in sub-Saharan Africa was initiated by the colonial governments of France, the United Kingdom and Belgium in the form of botanical garden networks co-ordinated by a central organization, such as the Kew Royal Botanic Gardens in the United Kingdom and the Colonial Gardens of Vincennes in France (Jain, 1990; Pardey *et al.*, 1991, 1995). These gardens concentrated on commercial crops that were either indigenous or introduced from other tropical regions. Among the major crops studied were: coffee, tea, oil palm, cocoa, cotton, groundnuts, rubber and sisal. The promising varieties of these crops, often screened and selected, were propagated and widely disseminated throughout the colonies. During this period, however, there was little varietal improvement.

In the 1900s, the botanical gardens were replaced by experimental stations which were established in most of the colonies under local colonial administration with technical backstopping from institutions in the region and in the colonial country. In the British colonies, adaptive research was carried out by the local colonial government and basic research was conducted in commodity- or discipline-based regional centres linked to an extensive international network. This system allowed for an active flow of germplasm and technology among the colonies in Africa as well as with other colonies outside the continent. The French system consisted of a number of commodity/ factor institutions which were based in France with a network of research stations scattered throughout the colonies. These institutions and stations carried out applied commodity-based research whereas the Office de Recherche Scientifique et Technique d'Outremer (ORSTOM) focused on basic research.

In 1933, Belgium established the Institut pour l'Étude Agronomique du Congo Belge (INEAC) in the Congo in order to serve the technical needs of their central African colonies (Pardey *et al.*, 1995). INEAC was one of the most extensive research networks in Africa with 36 stations governed by one central research station where research was carried out on food crops in addition to major export crops.

NARS in the early post-independence period

After independence, the responsibility for agricultural research in British east and southern Africa was transferred to the newly emerging governments. Some of the regional research institutions, such as the East African Agricultural and Forestry Research Organization (EAAFRO), operated as regional research organizations until the collapse of the East African Community in 1978 due to the divergent political and economic policies of the three member states: Kenya, Tanzania and Uganda. Despite this collapse, some of the commodity-based research foundations continued to function as effective semi-autonomous national institutions with funding from grower-paid levies (e.g. on coffee and tea). In Anglophone West Africa in the 1940s, Ghana was the leading centre for cocoa, Nigeria for oil palm and Sierra Leone for rice. After independence, these leading regional research institutions became part of the national system in each country.

Given the weak indigenous research system, the French maintained a strong presence in Francophone West Africa by providing management and substantial funding to agricultural research. These funds were provided through the Groupement d'Études et de Recherches pour le Développement de l'Agriculture Tropicale (GERDAT) followed, after 1984, by the Centre de Coopération Internationale en Recherche Agronomique pour le Développement (CIRAD) and ORSTOM, whose mandate spans beyond agricultural research. In the 1970s, most NARS in Francophone Africa became national responsibilities and continued to receive bilateral support from the two French international research institutions, CIRAD and ORSTOM.

As for the Belgian Congo research system, it fell into a state of disrepair after independence. This was largely due to the withdrawal of both Belgian financial support and expatriate research staff as a result of political upheaval and civil strife. The Belgian system had an excess of physical capacities including 26 research stations covering a total area of 300,000 ha. Following independence, INEAC was replaced by national institutes in the three former Belgian central African countries: Congo, Burundi and Rwanda. Although several donors had supported programmes under various research projects, the continued unrest and civil strife has since resulted in the destruction or weakening of these initiatives (Pardey *et al.,* 1991, 1995; Weijenberg *et al.,* 1995).

Recent Developments

Institutional developments

The NARS in Africa are currently dominated by the public sector in the form of a department within the Ministry of Agriculture or a semi-autonomous organization directly overseen by a ministry. These public institutions employ about 86.5% of full-time researchers and support staff, generally have an excessively large research infrastructure and are relatively overstaffed with an inadequate distribution of staff between disciplines. Several countries, such as Kenya, Malawi, Mauritius and South Africa, have semi-public autonomous agencies or foundations that carry out R&D on important export commodities such as coffee, tea, sugar and tobacco. These also provide funding for R&D through levies or local taxes on these commodities. These semi-public agencies employ about 3.5% of full-time equivalent research staff, whereas universities in most countries employ about 10%. Some countries, such as Nigeria and South Africa, have a relatively large number of faculties of agriculture or agricultural universities. However, university professors spend only about 15% of their time on research activities (Pardey *et al.,* 1997b).

Recently, there has been a transfer of responsibility from Ministries of Agriculture to one or more of the semi-autonomous national agricultural research institutes (NARIs), which cover either the entire agricultural sector or a subsector such as crops, livestock or natural resources. In countries such as Nigeria, responsibility for NARIs has moved back and forth from one ministry to another (agriculture, science and technology, education, etc.) thus exacerbating institutional instability (Odegbaro, 1984). In several sub-Saharan African (SSA) countries, these NARIs come directly under semi-autonomous research councils or have autonomous advisory and coordinating councils which give general and policy guidance. African NARIs and universities have also increased in size owing to:

- the availability in the 1960s and 1970s of government research funds;

- increasing contributions from donors to infrastructure and human resource development in the form of project 'enclaves' (whose relatively short duration is perhaps sufficient to build infrastructure and human capital but too brief to produce technology);
- government policies aimed at providing employment to new graduates;
- the lack of demand-driven strategic planning.

These poorly planned increases in the size of NARIs have impaired their ability to utilize resources efficiently as well as to manage their own long-term sustainability.

Most public semi-autonomous NARIs are currently trying to undertake reforms through: strategic planning and priority setting; downsizing the number of researchers and support staff (though not infrastructure); investigating sustainable funding alternatives; becoming more responsive to their clients; and improving their subregional, regional and international linkages to enhance their capacity to obtain, borrow and adapt technology.

In contrast with the NARIs, semi-public research agencies and foundations are relatively lean and efficient in allocating their resources and, in comparison, have not increased in size in terms of their staff and infrastructure. This institutional framework is quite sustainable because of their efficiency, responsiveness to their clients' needs and their track record in generating appropriate and profitable technology. They also benefit from the vertical integration of the agro-industries they serve such as tea, coffee, sugar and tobacco.

The Special Program for African Agricultural Research (SPAAR), in collaboration with African NARS, has been the prime facilitator in spearheading and crafting institutional developments and reforms to render national institutions in SSA more effective, participatory and demand-driven. SPAAR, through the Frameworks for Action (FFAs) to strengthen agricultural research (Spurling *et al.,* 1992; Weijenberg *et al.,* 1993, 1995; Taylor *et al.,* 1996) and in cooperation with the NARS and subregional organizations (where they existed), has helped to develop a common vision and agenda. This has been done in order to promote institutional innovations within national public institutions as well as to promote their linkage to all relevant stakeholders. This agenda is reflected in the principles indicated below.

Frameworks for Action Principles

- Institutionalize a strategic planning process that is participatory and responsive.

- Develop sustainable funding plans and mechanisms.
- Improve institutional and management capacity, transparency and accountability.
- Build country coalition and support groups involving those who produce, process, market, fund and consume.
- Strengthen researcher, extensionist, NGO, farmer and market agent linkages through refocused research agenda pertaining to on- and off-farm constraints.
- Promote regional and international collaboration (augment cost-effectiveness, spillover effects).

This collaboration has also involved the establishment and/or strengthening of subregional research organizations (SROs). Presently, the basic principles advocated in the FFAs are being implemented in all subregions in SSA and in a growing number of countries, for example, Benin, Cameroon, Côte d'Ivoire, Malawi, Mali, Mozambique, Tanzania, Togo and Zambia, where several of these principles are already incorporated in agricultural service and research projects funded by governments and donors. Illustrations of subregional and regional level progress include: (i) the establishment of both the Association for Strengthening Agricultural Research in Eastern and Central Africa (ASARECA) and the Forum for African Agricultural Research (FARA); and (ii) the evolution of CORAF (Conférence de Responsables de Recherche Agronomique en Afrique de l'Ouest et du Centre) as the subregional research organization for West and Central Africa. In recent years, the focus of SPAAR's programme has shifted from FFA development to the implementation of the FFA principles which result in strategic planning, sustainable financing, institutional reform, in-country and regional partnerships and participatory approaches to the generation and transfer of agricultural technology.

Human resources development

The number of agricultural research scientists in SSA increased approximately fourfold from 1961 to 1991 (Pardey *et al.*, 1997b). In addition, there has been a considerable increase in the number of technicians as well as support and administrative staff. The training of African scientists through expanded undergraduate university education, at the local level in most countries, and donor funding for postgraduate training abroad, have been responsible for this increase in the number of agricultural scientists. The high proportion of support and administrative staff in most NARS can be attributed to government policies to recruit all secondary school students and graduates and civil service regulations that have prevented downsizing once these policies were discontinued. The increase in the number of African research scientists corresponded with a decrease of expatriate researchers from 90% of the total number in the early 1960s to 11% in 1991, with Francophone countries having about three times more expatriates than Anglophone countries. In addition to the increase in the number of scientists, their qualifications have improved. Although the number of MS and PhDs has substantially increased, the proportion of scientists holding a PhD degree is unevenly distributed, with most in Nigeria and South Africa.

Despite these recent advances, human resources development is now confronted with the erosion of salaries and benefits and a general decline in operational resources. These two factors have resulted in significant brain drain of trained scientists, absenteeism or part-time research work, low morale and reduced productivity. However, attempts have been made to alleviate these constraints, including the provision of supplemental allowances and benefits from donor funding, the utilization of donor funds for operational expenditures and competitive contract research. Finding a long-term solution to combat the under-utilization and brain drain of available scientific staff is, indeed, one of the most important issues to be addressed in order to attain efficient and productive NARS in Africa.

Infrastructure development

Most African countries have acquired capacities in terms of the number and size of their research stations, experimental farms and the construction of laboratories, offices and residential houses. These developments have been greatly assisted by uncoordinated donor funding. However, the sites of several of these facilities have failed to attract high calibre research staff as their proximity to essential and other social services was not taken into consideration. Other facilities, inherited from former programmes, did not fully perform their research functions and yet drew on scarce budgetary resources. The fragmentation of NARS components (research institutes, extension services and universities) has also contributed to the inefficient utilization of research facilities. In turn, this fragmentation has hampered the optimal use of scientific human resources available within the NARS.

Agricultural Research Funding

The allocation of funds for agricultural research in Africa grew rapidly in the 1960s, moderately in the 1970s and, in general, stagnated in the

1980s in most countries. The availability of funds even declined in countries such as Nigeria where funds have decreased drastically since the mid-1980s (Pardey *et al.,* 1991). The decline of resources for agricultural research is a direct consequence of dwindling fiscal capacities of most transitional African countries currently undergoing structural adjustment programmes. Delayed government action to allocate available resources to priority sectors has forced a sparse distribution of already limited resources and has led to a general deterioration of all government-provided services.

The fourfold increase in the number of research scientists (especially during a period of reduced availability of financial resources) led to a sharp decrease in expenditures per scientist in most public research institutions throughout Africa. However, in semi-public institutions, such as commodity research foundations or boards, both the total allocation to agricultural research and the expenditure per scientist actually increased. These institutions have been funded mostly through levies or local taxes in addition to modest government and donor funding. Since they also tend to limit the number of scientific staff and maintain a high level of service, their scientists are generally more motivated.

Donor funding of African agricultural research increased from 34% of total research expenditures in 1986 to approximately 43% in 1991. With a few exceptions (South Africa, Nigeria, Botswana and Namibia), most NARS in Africa have developed an acute dependency on external assistance which, while useful in the short term, is not sustainable. There are already indications that donor contributions to agricultural research in Africa are decreasing. In the past, most of the donor support has gone to capital development and staff training; however, in recent years most donors have also funded operational expenditures. The decline in spending per scientist, in addition to the stagnation of agricultural research funding and the growth in the number of scientists, can generally be attributed to the cost of infra-structure development and maintenance; the high cost of staff training; a disproportionately high support staff to scientist ratio and costly expatriate technical assistance to staff certain NARS. Consequently, this pattern of resource allocation has resulted in higher spending on personnel; higher maintenance costs; poor conditions of service for research and support staff; and low morale, absenteeism and brain drain. Conversely, the semi-public institutions have avoided these negative consequences by remaining lean, efficient and competitive.

Above and beyond the decline of government allocation of funds to agricultural research in most countries in Africa, the flow of these limited resources is often erratic and untimely. Furthermore, the time-dependent nature of agricultural research, both seasonally and its overall duration, has also contributed to the inadequate output of

NARS in Africa. However, the generation and transfer of agricultural technology to farmers has been quite rapid by semi-public institutions and the private sector, often using production contracts. Examples of successful cases of technology transfer include: (i) the vertically integrated development of cotton in Francophone West Africa; (ii) the production of tea and horticultural crops in Kenya; (iii) tobacco and wheat in Zimbabwe; and (iv) sugar in Mauritius.

The Sustainable Financing Initiative

In light of declining funds from donors and inadequate government financial allocation, African countries, SPAAR and donors are forced to search for innovative ways to fund agricultural research using alternative and supplementary sources of funding. The Sustainable Financing Initiative (SFI), launched in 1995 by the Special Programme for African Agricultural Research (SPAAR) and the Multi-donor Secretariat (MDS) – with support from USAID and the World Bank – is one result of this search. Its objective is to assist African agricultural research and natural resource management institutions to identify, test and establish innovative means of funding their priority programmes. The SFI addresses the type and level of funding and its timeliness, stability in funding and financial accountability and the most appropriate use of funds (SPAAR Secretariat, 1996).

Several pilot programmes have been identified and initiated with African research and natural resource management institutions including: (i) The Council for Scientific and Industrial Research (CSIR), Ghana; (ii) the National Agricultural Research Organization (NARO), Uganda; (iii) the Network for Environmentally Sustainable Development in Africa (NESDA); and (iv) the Southern African Center for Cooperation in Agricultural and Natural Resources Research and Training (SACCAR). A second generation of SFI programmes has been launched in collaboration with the Kenyan Agricultural Research Institute (KARI), the Madagascar Environmental Foundation and the Association for the Strengthening of Agricultural Research in Eastern and Central Africa (ASARECA). These institutions had undertaken initial sustainable financing efforts on their own before collaboration with the SFI programme. In Kenya, these initial activities led KARI to cooperate with the private sector through the sale of seeds, payment for diagnostic and advisory services and contract research endeavours (Beyon and Mbogok, 1996).

Other initiatives are underway to target a more efficient utilization of existing resources through the institutionalization of strategic planning and priority setting and the introduction of demand-driven research involving client participation and competitive contract research (Office of Sustainable Development, 1996; Dunn, 1997; Kalaitzandonakes, 1997). The latter is designed to improve the quality of research and to encourage greater participation of all the NARS components in agricultural research (universities, the private sector and NGOs). One such example is the Contract Research Programme in Malawi, coordinated by the Agricultural Science Committee (ASC). Started in 1994, its projects involve university scientists and researchers with several public sector departments, particularly agriculture (SPAAR Newsletter, 1996).

Regional and International Research Efforts

The colonial regional collaborative efforts faded away in the early 1970s and were replaced by new initiatives from the Consultative Group on International Agricultural Research (CGIAR), as well as by support from the agricultural research institutions (ARIs) from the industrialized world. These institutions often used collaborative regional agricultural networks. Currently, over 20 subregional or Africa-wide networks are being supported and backstopped by CGIAR centres (Powell, 1995; Weijenberg *et al.,* 1995; Mullenaux *et al.,* 1996). Four centres (IITA, ILRI, WARDA and ICRAF) are based and active in Africa and another 12 have regional offices or substantial programmes in SSA. Their activities involve technology generation, the transfer of improved germplasm, training of scientists and the improvement of agricultural research management. In SSA there are also other collaborative programmes involving northern universities and research institutions.

For the past decade, in addition to CGIAR centre-coordinated networks, West African NARS – in collaboration with CORAF – have developed research networks that are coordinated by national scientists from one of the participating NARS. This coordination has been based on the comparative advantage and existing or potential capacities in the form of human resources, research and training facilities and scientific leadership in commodities or thematic research. Of late, most of the collaborative regional programmes executed by the international centres are being coordinated by subregional research organizations (ASARECA, CORAF, SACCAR) and their NARS. Along with support from their international partners, the increase in number and the improvement of African scientists' qualifications has paved the way for these scientists to play a more dominant role in research networking. International partnerships are presently undergoing a process of reorientation and renewal aimed at a more ecoregional approach which would integrate traditional commodities and factor research. The objective of this approach is to achieve high productivity and sustainable natural resource management.

Impact of Agricultural Research

Although the impact of agricultural research in Africa has not been as dramatic or substantial as that of the Green Revolution in Asia, there have been substantial achievements in many African countries at the farm level and in diverse agroecological zones. Collaboration between African NARS, IARCs and other advanced country institutions has facilitated:

- the availability of improved germplasm for many major crops;
- the use of appropriate integrated pest management;
- improved cultural practices for a number of important crops;
- improvements in animal production and health in several agro-ecological zones;
- successful technology transfer for traditional and non-traditional export crops such as horticultural food crops and ornamentals.

Rate-of-return studies (most were conducted on food crops with the exception of cotton in Senegal) in Africa (SPAAR Secretariat, 1996; Oehmke, 1997) indicate that agricultural research is a worthwhile investment. Twenty-one *ex-post* rate-of-return studies for agricultural research in Africa are summarized in Table 9.1. With a few exceptions, the rates of return are quite high and are more attractive than most other investments in Africa. These positive results have been achieved despite a difficult and inappropriate policy environment. Currently, there are indications that current economic and political reforms in Africa will produce even higher rates of return. Although there are no rate-of-return studies for research investment by the semi-public research agencies, companies and foundations, their survival and that of the industries they serve (without much government or donor support) are indications of the profitability of the technologies they generate. Three examples of successful research investment initiatives are discussed below.

Tea in Kenya

The area under tea increased from 21,448 ha in 1963 to 100,000 ha by 1993 with large estates constituting 31,300 ha and producing 99,374 tonnes, and smallholders representing 68,700 ha and producing 112,059 tonnes. The production of the large estates increased nearly six times in three decades while area expansion only increased 1.7 times, indicating that most of the increase came from higher yields. The increase in production from smallholders was from 312 tonnes in 1963 to 112,059 tonnes in 1993 and the area expanded from 3527 ha in 1963 to 68,700 ha in 1993. The large estate and the smallholders

Table 9.1. Summary of *ex-post* rate-of-return studies for African agricultural technology.

Author(s) and date of study	Location, commodity and years covered	ROR (%)	Comments
Makau, 1984	Kenya, wheat, 1922–1980	33	Econometric methods
Evenson, 1987	Africa, maize and staple crops	30–40	Aggregate RORs by region. Econometric.
Karanja, 1990	Kenya, maize, 1955–1988	40–60	Econometric. Statistical separation of research from extension; seeds
Mazzucato, 1992	Kenya, maize, 1978	58–60	Using Karanja data, finds minimal effect of fertilizer policy on ROR
Ewell, 1992	East Africa, potato, 1978–1991	91	Regional network/NARS collaboration
Sterns and Bernsten, 1992	Cameroon, cowpea, 1979–1991; sorghum, 1979–1991	3	ROR to research and extension
Schwartz *et al.*, 1993	Senegal, cowpea, 1980–1985	32–92	ROR to research-based famine relief includes all aspects of technology development and transfer (TDT)
Mazzucato and Ly, 1994	Niger, 1975–1991, cowpea, millet and sorghum	< 0	Non-adoption of varieties released in the study period
Laker-Ojok, 1994	Uganda, 1985–1991, sunflower, cowpea, soybean	135	6-year study period used due to civil unrest during the previous 15 years
Boughton and de Frahan, 1994	Mali, maize, 1969–1991	135	Introduction of maize into cotton system by CMDT
Sanders, 1994	Ghana, maize, 1968–1992, Cameroon, sorghum, 1980–1992	74	New cultivars with additional inorganic fertilizer. One new cultivar (S-35)
Smale and Heisey, 1994	Malawi, maize, 1957–1992	4–7	Improved research performance since 1985
Kupfuma, 1994	Zimbabwe, maize, 1932–1990	44	Research and extension activities of the Department of Research and Specialist Services (DRSS)
Khatri *et al.*, 1995	South Africa, aggregate agriculture	44	Econometric decomposition of agricultural productivity growth

continued

Table 9.1　continued

Author(s) and date of study	Location, commodity and years covered	ROR (%)	Comments
Ahmed et al., 1995	Sudan, sorghum, 1979–1992	53–97	New cultivar (HD-1) introduced in irrigation scheme, with additional inorganic fertilizer and erosion control
Quedraogo et al., 1996	Burkina Faso, maize, 1982–1993	78	Varietal improvement
Quedraogo and Illy, 1996	Burkina Faso, stone dykes, 1988–1994	7	Natural resource management/soil and water conservation technique
Seidi, 1996	Guinea Bissau, rice, 1980–1994	26	New cultivars for mangrove-swamp areas
Makanda and Oehmke, 1996	Kenya, wheat, 1921–1990	0–12	Based on econometric estimation of research impact on average yield
Armnade et al., 1996	South Africa, aggregate, 1947–1992	58	Econometric estimation, accounts for adjustment in the capital stock
Akgüngör et al., 1996	Kenya, wheat, 1921–1990	14–30	Time-series approach to econometric estimation of research impacts

Source: Oehmke (1997).

contributed 98.3% and 1.7% of production in 1963, however, in 1993 their contributions were 47% and 53%, respectively. The yield of tea in the large estates increased from 0.9 t ha^{-1} in 1963 to 3.2 t ha^{-1} in 1993, a threefold increase. For the smallholders, yields increased from 88 kg ha^{-1} in 1963 to 1.8 t ha^{-1} in 1993, a 20-fold increase. The national average yield, for the same period, increased from 0.8 t ha^{-1} to 2.3 t ha^{-1}.

As a result, the Kenya tea industry is a major success story. As an earner of foreign exchange, tea exports in Kenya are second only to tourism. In 1993, Kenya produced a record 211,433 tonnes of tea, exported 188,494 tonnes valued at US$318 million, and consumed 23,000 tonnes. This achievement is attributed to: (i) higher return on investment; (ii) good physical conditions for tea growing, i.e. rainfed, high altitude areas with a humid climate; (iii) favourable institutional setup consisting of large-scale tea company estates and smallholders supported by the Kenya Tea Development Authority with input supply, processing and marketing; and (iv) a sustained flow of technical innovation from the grower-supported Tea Research Foundation of Kenya (Noor, 1996).

Cotton in West Africa

The collaborative efforts of national cotton companies, the NARIs and their external technological and commercial links had a positive impact on the cotton industry in Francophone West Africa. The cotton industry and associated coarse grain crops (maize and, to a lesser extent, sorghum) saw rapid extensive development in the countries of Francophone West and Central Africa: Benin, Burkina Faso, Cameroon, Central African Republic, Chad, Côte d'Ivoire, Mali and Senegal. Cotton production, over the entire zone, doubled in the first decade after independence, and has more than tripled over the past 25 years. The region's share of world cotton exports increased from 4% to 9% between 1979–1981 and 1992–1993 (Bosc and Freud, 1994).

The growth in output was associated with impressive productivity gains. Yield increased from under 500 kg ha^{-1} to about 1100 kg ha^{-1} in the past 25 years. In some countries, yields reached over 1300 kg ha^{-1}. These yields compare well with the yields of rainfed cotton worldwide and are well above those in most sub-Saharan African countries. Fibre extraction rates have also increased from 35% in the early 1960s to over 40% in 1988. The area of maize cultivated in the cotton zone of West and Central African countries increased considerably owing to the diffusion of technologies for cotton which were also applicable to maize (Bosc and Freud, 1994). In Mali, the area under maize increased from 20,000 ha in 1980 to over 100,000 ha by the early 1990s. In Senegal, over the same period, the maize area rose from 6000 ha to 18,000 ha. In northern Cameroon, it grew from 7000 ha in 1982 to nearly 35,000 ha and in the cotton zone in Côte d'Ivoire, it more than doubled from 40,000 ha to 90,000 ha (Bosc and Freud, 1994).

Coffee in Uganda

The export of coffee in Uganda has experienced a fourfold increase in the last 6 years. This increase can be attributed to the liberalization of markets, which reduced smuggling, favoured external markets, and facilitated the introduction and adoption of high-yielding clonal varieties.

Outlook on Agricultural Research: Summary and Conclusions

A growing number of assessment studies have recently been carried out on the status of NARS and their ability to generate and transfer appropriate agricultural technology to their clients. Similar evaluations have also been done on collaborative regional and international agricultural research programmes targeting Africa, and to assess their impact and relevance. As a result of these studies, priorities have been determined and national, regional and international strategies have been prepared or are in a formulation stage. These assessments and various impact studies have also provided a wealth of information pertaining to agricultural production constraints and their likely solutions.

To meet the increased food needs of the next century, research must meet the challenge, in Africa, of ensuring that agricultural production increases by 4% annually. However, since the 1980s, the quantitative increase in scientists has not corresponded with qualitative improvements such as the provision of research services and generation of appropriate technology. The studies do, however, provide lessons, guidelines and recommendations on what should be done, and which organizations have the obligation to implement these changes. Elements necessary for successful technology generation, transfer and adoption include: a favourable policy environment and political support; a subsector approach to commodities; favourable credit and input supply; and participatory, responsive and closely linked research and extension. The prescriptions, mostly based on past successes, are given below. They apply primarily to the NARS and relevant governments of African countries as well as to those organizations (SROs, IARCs, donors and private sector companies) that operate at regional and international levels.

Based on this information, African NARS should strive to:

- ground agricultural research in the macroeconomic and sectoral policy of their respective countries;
- improve socioeconomic research capacities;
- institutionalize participatory strategic planning;
- develop strategic alliances with partners and clients (farmer associations, the private sector, NGOs, regional and international research organizations);
- partition agricultural research work between public- and private-sector research institutions;
- encourage the private sector to operate where commercialization of agricultural research is feasible;

- learn how to craft appropriate and cost-effective organizations from similar foundations, thus to optimize overall resource utilization; improve the condition of service for research staff, maximize their contribution to agricultural research, improve morale and reduce the loss of trained scientists (brain drain);
- advocate the enhancement of local postgraduate training at universities;
- achieve and diversify means of sustainable financing;
- improve the coordination of donor assistance and reduce dependency on external funding;
- make agricultural research more demand-driven and participatory;
- facilitate the flow and transfer of technology, including new varieties of seeds, from regional and international (public and private) institutions.

At the regional and international levels, there is a clear need for:

- preparation and implementation of strategic plans for subregional collaborative agricultural research programmes;
- an effective information and technology flow within each subregion and, in time, between subregions;
- institutionalizing regional and international collaboration in agricultural research by incorporating these centres into the plans and budgets of national programmes;
- providing assistance to NARS to develop their capacity to obtain, test and adapt agricultural technology;
- preparation of subregional plans for human resource development to include in-service, advanced degree and leadership training within the subregion;
- the development of long-term international partnerships with public and private sector agencies.

Morocco

10

Ali Kissi and Alia Reguragui

Institut National de la Recherche Agronomique, Rabat, Morocco

Introduction

Morocco's system of agricultural research is one of the African continent's oldest, dating from the early 1900s. The system has evolved with the agricultural development strategies adopted by the country. The system now faces numerous challenges as a result of national and international developments, and the goal of achieving sustainable and balanced agricultural development throughout the different regions of the country. Recent developments that have affected Moroccan agriculture include:

- The 1992 Rio Conference on Environment and Development, which brought the concept of sustainable agricultural development to the fore.
- Competition at various levels as a result of GATT (now WTO) policies.
- Efforts to achieve greater integration between the international system of research and national systems as a way to globalize agricultural research at the national regional and international levels.
- Donors placing a higher priority on cooperation with regional rather than national organizations.

At the national level, the main developments include the following:

- State divestiture and disengagement from direct actions carried out in the past to benefit farmers.

- Role that the region is now expected to fill in political, economic and social affairs.
- Farmer involvement in development projects at the local (commune) and regional levels.
- Reduction of public resources allocated to research and dwindling resources from international cooperation.

All these changes have affected, and will continue to affect, the development of Moroccan agricultural research. Before examining how the system has responded to these changes, first we will provide an overview of the organizations involved in research and the way in which the system operates.

Organizations Involved in Research

Agricultural research in Morocco encompasses the public sector, which has traditionally played a major role, and the emerging private sector. Numerous public sector groups are involved, including institutions primarily engaged in research, institutions involved in education and research, and R&D groups.

Research institutions

The National Institute for Agricultural Research (Institut National de la Recherche Agronomique, INRA) and the National Centre for Forest Research (Centre National de Recherche Forestière, CNRF) are primarily engaged in research. INRA was created in the 1920s. It is responsible for scientific, technical and economic research to improve agriculture. It also carries out forward-looking studies, particularly on the natural environment and improvements in plant and animal production. INRA undertakes, on its own initiative or at the request of private individuals, trials on new and improved crops and animal production and experimental work on the farming, processing and utilization of plant and animal products.

INRA is responsible for disseminating information on its own research, as well as research conducted in other countries; studying and recommending practical methods for applying research results; and providing advice to organizations involved in agricultural extension and to farmers themselves. INRA is also responsible for commercializing the results of research and monitoring research being conducted on behalf of public agencies. To achieve its goals, INRA carries out research on farmers' fields and at 24 experimental farms.

These activities are grouped into 16 programmes supported by seven departments: Agronomy, Plant Breeding, Plant Health, Socioeconomics, Animal Husbandry, Physical Habitat and Technology. INRA relies on the Division of Information and Training and on the Research Development Units (Services de Recherche et de Développement, SRD) set up within the Regional Centres for Agricultural Research (Centres Regionaux de Recherche Agronomique, CRRA) to disseminate the results of its research.

CNRF was established in the 1930s, with four departments: Tree Farming, Genetic Improvement of Forest Species, Wood Technology and Ecology. CNRF conducts scientific, technical and economic studies which promote conservation and sustainable development of the forest and its resources. It also carries out innovative studies on the natural environment and forest plant life, in cooperation with training and research institutions at the national and international levels. CNRF also plays a role in documentation and dissemination within its areas of expertise. One of its responsibilities is to coordinate the actions of the Division of Water, Forests and Soil Conservation in documenting its own research, as well as research performed in foreign countries, and disseminate the results.

Institutions involved in education and research

There are three institutions involved in education and research: the Hassan II Institute of Agronomy and Veterinary Science (Institut Agronomique et Veterinaire, IAV Hassan II), the National School of Agriculture in Meknes (École Nationale d'Agriculture, ENA de Meknes) and the National School of Forestry in Sale (École Nationale Forestière d'Ingenieurs de Sale, ENFI). In addition, some university science departments are also involved in research. ENA de Meknes, founded in 1942, is the oldest institution devoted to education and research. It provides:

1. Training. ENA trains agronomists who are specialists in the subjects most relevant to the agricultural development requirements of the country.
2. Research and development. Numerous departments are involved in R&D activities, such as agronomy and plant breeding, arboriculture and viticulture, plant ecology, rural economy, rural infrastructure, agricultural mechanization, plant pathology, animal production and pastoralism, soil sciences, food technology and zoology.
3. Transfer of knowledge and skills. ENA is also involved in transferring the knowledge and skills acquired by its teacher–researchers, through the National Centre for Extension Studies and Research (Centre

National d'Études et de Recherche en Vulgarisation, CNERV), which is located within the institution.

IAV Hassan II, established in Rabat in 1966, includes a training and development complex in Agadir that specializes in horticulture. It provides: (i) training: the Institute's teacher–researchers provide training in agronomy, veterinary medicine, agro-based food production, water engineering, topography, agricultural mechanization, forestry, wood technology, horticulture and plant health, land use planning and fishing technology; (ii) research on development issues: research is conducted in real settings and at the Institute's three experimental farms.

The primary mission of ENFI, which was created in 1968, is to train forestry engineers. ENFI offers advanced social, economic, technical and scientific training, chiefly in subjects related to managing and developing the nation's forests and natural resources, as well as promoting economic development in mountainous regions. Apart from its teaching and training function, ENFI also conducts studies and research, at the request of government agencies and private individuals, on the nation's forests and natural resources and the economy of the mountainous areas.

Some university science departments, particularly those located in agricultural areas, are also becoming more and more involved in agricultural research. This is especially true of the science departments in Fez, Meknes, Kenitra, Settat, Marrakech, Agadir, Beni Mellal and Oujda. This is a recent development related to their institutional growth and the socioeconomic role that they are expected to fulfil. They have, however, limited resources and their objectives are not specifically based on agricultural development strategies.

Thus, considering the different orientations of the organizations involved in the national system of agricultural research, INRA's key partners are IAV Hassan II, ENA de Meknes and the various science departments. CNRF's key partners are ENFI and the science departments.

Research and development agencies

The R&D agencies at the central level are the Official Laboratory for Chemical Research (Laboratoire Officiel de Recherches Chimiques, LORC), the Agricultural Water Systems Experimentation Unit (Service des Experimentations d'Hydraulique Agricole, SEHA), and the Experimentation, Trials and Standardization Unit (Service des Experimentations, des Essais et de la Normalisation, SEEN). At the regional level, development agencies such as the Regional Offices for Agricultural Development (Offices Regionaux de Mise en Valeur Agricole, ORMVAs) and the Livestock Directorate maintain adaptive research units. Most research and R&D institutions have experimental stations to support their research activities.

Network of research stations within the system

The network is fairly complex and is dominated by the INRA stations as follows:

- INRA has 24 experimental farms, with about 4800 ha located in different agricultural regions of the country.
- CNRF has an experimental network of 150 experimentation sites and 30 arboretums. CNRF also maintains a regional centre in Marrakech for forest research in arid zones.
- ENA de Meknes has only one experimental farm, adjacent to the school.
- IAV Hassan II has three research stations, two specializing in stock raising in Gharb and Tadla and one specializing in horticulture in Agadir.

These stations are unevenly distributed around the country, with most concentrated in the central and western regions. Half of the stations were created to respond to the development needs of newly established irrigation areas (Table 10.1).

In addition to research stations, the country has a certain number of R&D units. For example, the ORMVAs (the regional offices responsible for providing extension services to farmers involved in irrigation schemes) maintain R&D facilities referred to as Experimentation and Agricultural Development Stations (Stations Experimentales et de Mise en Valeur Agricole, SEMVAs) where verification and demonstration trials are conducted. The Gharb ORMVA maintains a specialized research facility, the Sugar Cane Technical Centre, which plays a major role in developing this crop.

Similarly, the Livestock Directorate has a network of experimental farms for animal breeding and research on herd management (Gouttitir, Ouled Isly and Ain Khettar in the eastern region; Touna in the mid-Atlas region; and Krara in the Chaouia region). The Livestock Directorate also carries out research on indigenous forage and range species at the Centre for the Production of Range Seeds (Centre de Production des Semences Pastorales, CPSP) in the Doukkala region. Some of the ORMVAs are planning to close down certain SEMVAs because of funding constraints.

In summary, the public sector's role in agricultural research is characterized by the diversity of its constituent parts. However, while

Table 10.1.   Principal research stations by ecosystem and agroecological region.

Ecosystems	Research stations
Irrigated agriculture	
Basse Moulouya	INRA: Bou Areg
Gharb	Allal Tazi-El Menzeh
	IAV: Moghrane
Loukkos	INRA: Larache
Tadla	Afourer-Deroua
	IAV: Tadla
Doukkala	INRA: Khemiss Zmamra
Haouz	Menara-Tessaout
Souss-Massa	Melk Zhar
Oasis	INRA: Zagora-Errachidia
Continental rainfed agriculture	INRA: Douyet-Merchouch-Ain Taoujtate
	ENA: Hadj Kaddour
North Atlantic rainfed agriculture	INRA: Boukhalef
Mid-Atlantic rainfed agriculture	INRA: El Koudia-Test garden El Guich
Semi-arid conditions	INRA: Jemma Shalin Sidi El Aidt Jemaa Rhiyah
Saharan conditions	INRA: Laayoune
Mountainous conditions	INRA: Annoceur

this diversity significantly enhances the national system, it also requires considerable coordination.

Private sector

In Morocco, the private sector's involvement in agricultural research is limited to a small number of groups engaged in adaptive research. The most significant players in the private sector are: (i) the Moroccan Agricultural Services Company (Société de Services Agricoles Morocaine, SASMA), a long-standing enterprise that is an outgrowth of the citrus and vegetable subsectors; and (ii) the Interprofessional Technical Centre for Oilseeds (Centre Technique Interprofessionnel de Oleagineux, CETIO), established quite recently (1994) and linked to the oilseeds subsector. In addition to these formally structured organizations, there are seed companies and companies that market fertilizers and pest and disease control products. They engage in adaptive research or verification trials, sometimes in collaboration with research institutions from the public sector, to enhance their marketing activities. Producer associations (beet producers, breeders, etc.) are also involved in some adaptive research activities. This trend is likely to increase, and some associations have every intention of playing an active research role in matters pertinent to their particular sector. One example of this trend is the Association of Fruit and Vegetable Producers and Exporters, which plans to create an agricultural network in the rich farming region of Souss, by establishing partnerships with the various agencies involved in R&D in the region.

Some particularly innovative private farmers are also involved in adaptive research as a way to promote their own agricultural activities. The same is true of certain state-owned agricultural enterprises, such as the Agricultural Development Company (Société de Développement Agricole, SODEA) and the Agricultural Lands Management Company (Société de Gestion des Terres Agricoles, SOGETA). These activities are sometimes carried out jointly with INRA, IAV Hassan II, or ENA de Meknes and the results obtained are often made available to the larger community of farmers.

Coordinating bodies for agencies that fall under MAMVA

The main coordinating bodies are:

- DERD (Directorate of Education, Research and Development/ Direction de l'Enseignement, de la Recherche et du Développement): located within the Ministry of Agriculture and Agricultural Development (Ministère de l'Agriculture et de la Mise en Valeur Agricole, MAMVA), DERD is responsible for coordinating agricultural research and, more specifically, orientating, coordinating and monitoring educational, research and extension activities.
- GIRA (Interprofessional Agricultural Research Group/Groupement Interprofessionnel de la Recherche Agronomique) was recently created. Its membership is composed of research managers and scientists. GIRA is a consultative body which could play an important role in agricultural research.

The National Centre for Coordinating and Planning Scientific and Technical Research (Centre National de Coordination et de Planification de la Recherche Scientifique et Technique, CNCPRST) has been in existence since 1975. CNCPRST falls under the administrative supervision of the Ministry of Higher Education, Professional Training and Scientific Research, and is charged with developing, orientating and coordinating scientific and technical research of all types.

However, it has not yet played a role in agricultural research. Thus, there is no specific body that coordinates Moroccan agricultural research in its entirety. Unlike many other countries, Morocco has no national council on agricultural research.

In conclusion, it would seem relevant to cite a passage from the 'Higher Education and Research' chapter of an evaluation report on DERD management, drafted by the General Development Council (a MAMVA body):

> Growth in the system of research and higher education reflects the weight of a vertical and hierarchical administrative structure. Each government service expects to carry out its own research and train its own managers. When a new need arises, the typical reaction is to create a new structure rather than diversify the scope of existing structures. The growth of the system is discontinuous. This type of growth, characteristic of a hierarchical and centralized business culture, has a natural tendency to marginalize the role of any consultative or coordinating body.

Organization

Administration and statutes

State bodies
State bodies are administered by government authorities and are subject to public accounting regulations. With the exception of university science faculties, which answer to the Ministry of Higher Education and Executive Training, all public sector units answer to MAMVA. Although they answer to the same body, the units are governed by different statutes:

- INRA, IAV, ENAM and LORC are public establishments, recognized as legal entities and accorded financial autonomy. Administered by the state and subject to state financial control, they are managed by an administrative council led by the Minister of MAMVA. The composition of the council varies from one body to another.
- ENFI is part of a central directorate of MAMVA.
- CNRF reports to the Directorate of Water, Forests and Soil Conservation of MAMVA.
- The Service for Experimentation, Testing and Standardization and the Service for Agricultural Hydraulic Testing report to the Rural Engineering Administration, which is also part of MAMVA.

Private societies

Private societies are managed by administrative councils on which all facets of the branch are represented. They are subject to private accounting regulations. These disparate statutes and the absence of links make it difficult to share resources, in particular human resources, all the more so since the career structures vary. This is a major handicap to the consolidation of a truly national system of agricultural research. Indeed the statutes of the public establishments do nothing to encourage staff mobility, since each individual post is budgeted for, or dispensed with, as in the civil service. Public research establishments, although enjoying financial autonomy, are subject to strict public accounting regulations when they put their budget into practice: budget estimates, resource allocation, commitments and execution. This rigid budget management and financial control precludes any transfer of funds. The statutes of public research establishments should thus be revised, facilitating the mobility of human resources and providing for more flexible budget management, in order to consolidate a national research system.

Human resources

It is difficult to determine accurately the number of scientists involved in national agricultural research. A study revealed the following figures: scientists employed by research institutes (INRA and CNRF) each spend 70% of their time on genuine research work, while the figure for teacher–researchers is 30%, and for engineers employed by the R&D units it is 40%. These averages are only indicative. The total is equivalent to approximately 450 research scientists working in the state agricultural research sector. To this figure we must add the equivalent of 20 research scientists working in the private sector. The overall potential is thus some 470 research scientists, which makes the Moroccan national system one of the largest on the African continent.

The level of training of research scientists has risen significantly, with 60% of research scientists and teacher-researchers holding a PhD or equivalent. This percentage is higher in the teaching and research bodies than at INRA. Although 70% of IAV research staff, and almost 50% of ENAM and ENFI researchers hold PhDs, only about 40% of INRA researchers do. These numbers are likely to increase.

Funding

Funding public research

The main source of funding for public research remains the state, with support from bilateral and multilateral funding agencies. The *state budget* for agricultural research is subject to certain restrictions every year. If we take the example of INRA, the total budget of the Institute has dropped in nominal and in real terms (in dirhams taking an average rate of inflation of 5%) since 1993. Over the past 4 years INRA's budget has thus been cut on average by 2–6% in nominal terms. In real terms the drop has been about 6.6%. Over the same period, the per capita funds allocated to procure equipment and inputs, in terms of the number of research scientists employed in the base year (233 research scientists in 1993), have dropped on average by 10% in nominal and 13% in real terms. The actual per capita drop is, in fact, even greater, since the number of research scientists rose by some 10% from 1993 to 1996, making the actual drop around 12% in nominal and 15% in real terms. Given that certain general items of expenditure have not been affected by cuts, the drop in funding has hit research work particularly hard.

Support accorded by *bilateral cooperation funds* over recent years from the USA, the former Federal Republic of Germany, France, Canada and Italy, have been extremely significant. INRA and the teaching and research establishments have benefited from many cooperation projects, some of which were quite large. The funds attached to these projects, when added to those allocated by the Moroccan state, have made it possible to set up infrastructure, train research scientists and fund research activities. Bilateral support is, however, declining. The large-scale projects of the 1980s have given way to more modest projects, aiming in particular to establish closer links (e.g. cooperation with France and with the USA), or to enhance management procedures in research and the ways of applying research results (e.g. cooperation with Germany).

Funding private research

Private research is fully funded by the profession that it serves. Thus the budget of SASMA (citrus and vegetable subsector) comes from levies on exports of plums and early agricultural produce. CETIO (oilseeds subsector) is financed by contributions made by professional families in line with the number of seats they hold on the administrative council. Government authorities also support some private research units by providing research staff.

Reactions of the National System to Changes

Recent national and international events present challenges, primarily the search for additional resources from the state, determining the optimum way to use available resources at national and Maghreb levels and an appropriate division of responsibility between the state and private research sectors. How has the national system reacted to these challenges?

Search for additional resources

Faced with cuts in the state budget, combined with a drop in outside funding, research managers and scientists are attempting to find additional sources of revenue. Two channels have so far been explored: contract research work and 'own revenue'.

Contract research work
At INRA 89 R&D contracts have been received since 1986 by various departments, 71 of these within the past 5 years. The number of research contracts has been falling drastically, with only five in 1997. The contracts are with individual farmers, farmers' groups, state development bodies and input production or marketing associations, some of the latter being state run. The contracts for state development bodies accounted for 37 of the 89, with associations and individual farmers next at 25 and 23, respectively. There is a heavy regional bias in the way the various regional agricultural research centres benefit from this source of funding. Two of the eight centres attracted more than half of the 89 contracts.

At Hassan II Institute of Agronomy and Veterinary Science, contract work was started early, and covers research, training and services. They had 493 national and international contracts between 1972 and 1995, with a drastic drop recorded in 1995. The drop at INRA might partly reflect a lack of interest of research staff in attracting new contracts. This is due also to reduced funding and to the centralized management of contracts, which results in delays in procurements and a lack of material incentives for staff to take on the additional burden. For this reason, the management of INRA is negotiating with its administrators in an attempt to achieve more flexible management of research contracts and a system of incentives.

Contract research brings in significant additional income to some components of the national system. It must, however, be managed with care and should focus primarily on priority research topics or on the application of results achieved. A detailed plan should be drawn up regarding the uses to which the revenue generated is put, including incentives for research staff and, if appropriate, their collaborators. In other words this form of research should follow a new, well thought through strategy, serving the general interest rather than specialized interests and, above all, not deviating from established research priorities. Recently the government entered into an agreement with representatives of the agricultural sector to promote national agriculture. One provision of this agreement covering agricultural research stipulates that:

- The government shall undertake to strengthen agricultural research in partnership with professional organizations, on the basis of a common programme in line with the topics of interest to the various branches.
- Producers in turn shall undertake to contribute to the funding of these training and research activities.

Own revenue

At INRA additional revenues come primarily from marketing 'farm' products, i.e. seeds and royalties paid for the production of new varieties. These revenues account for most of INRA's self-funding efforts, which make up between 8 and 13% of the Institute's budget. This policy does not find favour with the research staff as a whole, some of whom see this as a waste of the Institute's resources which they feel should be dedicated to the priority task of research. As for the teaching and research establishments, own revenues are generated primarily by services in terms of water and soil analyses, diagnosis of disease, veterinary care and technical and economic consulting services. These establishments also have special units to provide these services. As is the case with contract work there are legal limitations to the boosting of own revenues, and they must be managed with care.

To sum up, new additional resources might help offset cuts in public funding and outside funds, but must be accompanied by better management of the funds allocated by the state, in particular a reduction in overheads, which would increase the percentage of funding available for genuine research work.

Optimizing use of resources

Like all scientific research, agricultural research is a costly undertaking. Research staff must be trained, infrastructures set up, highly sophisticated equipment procured, and funds found to operate the system. Within the components of the national and Maghreb system there are top-class research scientists, infrastructure and modern equipment. One of the major challenges, however, is to optimize the way these resources are used. To this end, several activities have been conducted: developing coordination and seeking to understand better the needs of clients at the national level; and establishing research networks at the Maghreb level.

Developing coordination
The process of objectives-oriented planning based on joint identification of problems and priorities holds promise. This process should lead to national (rather than institutional) programmes to be implemented jointly by the research staff employed by various components of the national system, spanning the public and private sectors. This process attempted to achieve common utilization of the resources of each component, with a view to solving the problems identified together. Unfortunately, this process, which worked fairly well during the planning phase, has not been continued during the implementation phase.

Parallel to objectives-oriented planning, other tests have been run, with the encouragement of national and foreign funding bodies, with a view to setting up operational coordination mechanisms. Of these tests the following are particularly worthy of comment:

- The implementation of federal projects developed by the research staff of INRA, the Agricultural Faculty of Hassan II and the Meknes ENA with funds from DERD.
- The reshuffling of the responsibilities of these three establishments to resolve the priority problems facing agricultural development with the help of applied research projects. These are projects to support agricultural development in irrigated areas (PDSA) and projects to develop *bour* regions (PMVB).
- The development of relationships between research scientists and these establishments and their counterparts in France through research projects for agricultural development (PRDA).

These are effective units, but their activities are not well coordinated and they sometimes compete with one another. This situation developed when there were adequate funds which allowed each unit to act independently. It is also the result of the attitudes of certain funding bodies which dealt separately with each component of the system. Nevertheless, the essential reason for this lack of coordination is the difficulty of establishing appropriate coordination mechanisms. This difficulty is in turn the result of several factors:

- The diversity of the statutes of public units even among financially autonomous bodies, which hampers the transfer of funds from one body to another.

- The public bodies answering to MAMVA report to different directorates, each of which has its own concerns.
- Most public bodies have developed their own culture and in some cases their own specific clientele.
- A lack of strong political will.

Until recently, the problems involved in coordinating national research only applied to the coordination among various public-sector components. With the strengthening of the private sector, however, this coordination must assume three dimensions: coordination among the public-sector components, among the private-sector bodies and throughout the entire national system. In other words, the coordination of national research is no longer the exclusive domain of the state. The private sector will also be involved, as will clients, who are becoming increasingly demanding and increasingly influential.

Better understanding clients' needs
If we are to optimize the way available resources are used, we must respond to real priority problems facing Moroccan agriculture. Today more than ever, the components of the national research system are subject to rising pressure from clients. Foreign competition, which is set to rise, and the challenges of contributing to sustained and region-ally balanced national agricultural development, make for a greater demand for new technologies, and come from a wider and more demanding clientele.

- Among tests run to identify the needs of clients, we should look at efforts to involve clients in the development process within the scope of objectives-oriented planning. This process provides for clients to participate in the studies and the various stages leading to the elaboration of the research programme for a given sector, and more particularly in those referring to the identification of constraints and the establishment of priorities.
- The other test consisted of creating GIRA. The aim was to set up a hybrid organization bringing together professionals and managers of national research, to direct research more effectively to the real priority needs of the various branches.

Maghreb research network
Agricultural development throughout Maghreb countries faces similar constraints, the most important being the climatic zone. These coun-tries also possess the same potentials in terms of the same ecosystems. In this context it is quite logical to mobilize the resources of the states involved to develop technologies likely to ensure harmonious agri-cultural development. Research and R&D networks supported by bilateral cooperation and international research centres have been set

up and have produced results, some of which have already had a significant impact on the agricultural development of these countries.

The major problem is how these networks can continue to survive once funding bodies discontinue assistance. On the basis of the experience of one such network, the RMLB (Reseau Maghrebin de Lutte contre le Bayoud), it is safe to say that this problem is acute. The activities of this network, extremely dynamic for as long as it received FAO financial assistance, have been on hold since the funding body discontinued support.

Sharing responsibility between sectors

The private sector has been expanding recently. In the past this sector was geared to applied research, whereas today it is also a producer of new varieties. This selection of more effective new varieties has hitherto been considered the domain of the public sector, because of the long lead-time involved. Now, with free international trade, every farmer or group of farmers can purchase material, study its behaviour under local conditions and submit the varieties selected to be included in the official list as new varieties. The private sector is now interested in varieties of strategic crops, such as cereals and food legumes. If this trend continues the public sector seems set to lose the privilege of being the only national producer of new varieties of strategic crops. The foreseeable rise in private research will have a significant impact on the national research system, in particular regarding the distribution of roles, funding, coordination and cooperation policy.

In some industrial countries public research institutes focus more on fundamental research and conduct applied research only in sectors that are not of interest to the private sector. Is this sort of development possible in Morocco? Could the private research sector develop enough to meet the needs of a wider range of farmers? It is becoming important for the public sector to push itself on the national research market taking into account its comparative advantages, and focusing its efforts accordingly. This is true in particular of the research institutes and teaching and research organizations. Research and development organizations would benefit from greater integration into the private sector. It is quite probable that in future the role of the private sector will be so decisive in some fields that the state will be called upon to support it more, even financially. Yet, this option, which might be justified by a cost–benefit calculation, must be based on a well thought through strategy and supported by the various partners.

Conclusion

The Moroccan agricultural research system has developed significantly over the past 30 years. Development of research staff in particular has been rapid in terms of numbers and qualifications. Within this system, the public sector is dominant, although some relevant units in the private sector have already proved their worth and play an important part in the development of the sector that funds them. Otherwise, alongside INRA, teaching and agricultural research establishments have formed the *avant-garde*, and continue to do so. University faculties of science in agricultural regions are called on to develop their capacities for agricultural research.

At present there is no real coordination between the various components of the Moroccan system. Tests have been conducted, but they must be consolidated to make the system genuinely operational. There is also a lack of any national level coordination body. The state sector is facing a number of challenges in the wake of changes at the national and international level. Its development will depend on its reaction to cuts in funds and on the future role that the private sector is called on to play. In view of the decline in resources, attempts have been made to develop contract research work, own revenues, and networking, especially at Maghreb level. The national research system needs a strategy laying out objectives and according specific roles to the various components of the system, in the state and private sectors. The guidelines for this strategy should be drawn up by a national agricultural research body and should involve the various branches of agriculture as partners. If the system is to operate and harness synergy, it also needs coordination mechanisms and flexible management, as well as links between the various statutes of the component parts of the system.

It is difficult to forecast the future of the Moroccan research system. All that we can say is that certain trends will become more marked, especially the rise of the private sector, the role of the university, the search for quality, regionalization, better coordination, networking and the involvement of clients in planning, funding, implementation and the transfer of results. The system is being called on in particular to adjust to the decline in available funds, by making institutional and structural changes. These changes will try to lighten the network of research stations and thus cut back on administrative costs. These changes will also be accompanied by the adoption of new mechanisms for managing research in terms of planning, monitoring, evaluation, and transfer of results and resources. Some of these changes will come up against old habits, established rights, and bureaucracy; but the state, in particular through the allocation of resources, can influence the changes. The same applies to financing bodies which will be called on to support new collaborators. Finally, the clients of research will become increasingly influential and will play a larger role in future.

Sector Studies

Principles for Public Investment in Agricultural and Natural Resources Research

11

Julian M. Alston[1] and Philip G. Pardey[2]

[1]*Department of Agricultural and Resource Economics, University of California, Davis, CA 95616, USA;*
[2]*International Food Policy Research Institute, Washington, DC 20036, USA*

Introduction

Significant changes are taking place in the financial support for agricultural R&D and in the roles played by national and international governments and their agencies, and the private sector. Some aspects of these changes represent a continuation of longer-run trends; others represent dramatic departures from previous patterns. Alston *et al.* (1998b) document in detail recent changes in agricultural research institutions and investments in five OECD countries (Australia, the Netherlands, New Zealand, the UK and the USA). A chapter in Alston *et al.* (Pardey *et al.*, 1998a) provides an overview for the OECD as a whole. These studies show that industrial countries have shifted markedly towards more private funding of agricultural R&D and increased private provision of agricultural R&D. They document that public sector roles have been changing in response to a slowing of the growth in public support for research in general, and agricultural research in particular, pressures for increased accountability for the use of public R&D funds, and a broadening of the agricultural research agenda to include issues such as food safety and quality. In addition, both private and public research agencies have had to adapt to deal with increased concerns about the environmental consequences of agriculture and of new agricultural technologies, and sizeable shifts in the basic biological sciences which underpin the agricultural sciences, in addition to a strengthening of the intellectual property protection afforded to the new crop varieties and other inventions.

©CAB INTERNATIONAL 1998. *Investment Strategies for Agriculture and Natural Resources* (ed. G.J. Persley)

Similar policy, institutional, scientific and research funding developments have occurred in developing countries, although the pace and pattern of change are far from uniform among countries. Notably, the rate of growth in public support for the agricultural sciences has slowed even more markedly in developing countries, and for many it has actually decreased. And, at the same time, support for the Consultative Group on International Agricultural Research (CGIAR) centres has been under pressure, and support for some key centres has shrunk. This widespread slowdown or contraction in the growth of support for public agricultural R&D has generated a demand from various quarters for policy recommendations and actions to deal with these developments. Some have called for a return to the past patterns of generally strong growth in public funding for publicly performed R&D. Others have sought an even greater private role in the funding and performance of R&D. New public–private partnerships have also been canvassed. Some governments have responded to these pressures and devised new policies and institutional arrangements for funding agricultural R&D. Others have gone further and redesigned the institutions that allocate research resources, and in some cases revamped the public agencies that perform the research. While some of the changes have occurred in a piecemeal, haphazard fashion, in some countries processes of change have been more orderly, involving rounds of serious policy analysis and debate.

Although political opportunism plays a role in all such public policy and institutional change, there is also a role for objective argument based on facts and reason. Changes in agricultural R&D policy are inevitable, and may be desirable. This chapter offers a statement of the economic arguments relevant to designing policies that will sustain public and private support for agricultural R&D, with a total investment and mix of R&D that is consistent with achieving the greatest net benefits to society as a whole. We spell out the economic principles that justify government involvement in agricultural R&D and should also guide the nature and extent of that involvement, and the linkages between the private and public sector roles. Specifically we briefly revisit the rationale for public involvement in agricultural R&D, draw on an economic framework to develop the policy principles that underpin this public role and then consider some of the options for government involvement in agricultural R&D. As background information, we begin with an overview of the global trends in public funding of agricultural R&D, and provide more specific information on the sources of support among industrial countries.

Investment Patterns

Global trends

Investments by national governments worldwide in public agricultural research almost doubled in real terms over 20 years, from $7.3 billion (1985 international dollars) in 1971 to nearly $15 billion in 1991 (Table 11.1).[1] Annual expenditures on publicly performed agricultural research in developing countries grew by 51% from $3 billion (1985 international dollars) in 1971 to $8 billion in 1991. Across the industrial countries, annual public agricultural spending grew by 2.3% from $4.3 billion (1985 international dollars) in 1971 to $6.9 billion in 1991 and $7.1 billion by 1993.

For all regions of the world, however, real R&D spending grew at a much slower pace during the 1980s than in the 1970s. In 1971, as a group, developing countries accounted for 59% of the spending. By 1991 the situation had changed markedly. Developing country R&D spending

Table 11.1. Public agricultural research expenditures, global trends.

	1971	1981	1991
Expenditures (millions 1985 international dollars)[b]			
Developing countries	2,985.2	5,534.8	8,016.7
Sub-Saharan Africa	699.2	927.2	968.4
China	457.4	939.4	1,494.3
Asia and Pacific (excl. China)	861.5	1,922.4	3,501.6
Latin America and Caribbean	507.9	1,007.7	950.7
West Asia and North Africa	459.2	738.1	1,101.7
Industrial countries	4,298.1	5,713.4	6,941.4
Total[a]	7,283.3	11,248.2	14,958.1
Average annual growth rates (%)			
Developing countries	6.4	3.8	5.1
Sub-Saharan Africa	2.5	0.8	1.6
China	7.7	4.7	6.3
Asia and Pacific (excl. China)	8.7	6.2	7.3
Latin America and Caribbean	7.2	−1.1	2.7
West Asia and North Africa	4.3	4.0	4.8
Industrial countries	2.7	1.7	2.3
Total[a]	4.4	2.8	3.6

Source: Pardey *et al.* (1998a).
[a] Excludes Cuba and former Soviet Union.
[b] To obtain an internationally comparable measure of the volume of resources used for research, research expenditures were compiled in local currency units, then deflated to base year 1985 with a local GDP deflator (World Bank, 1995a), and finally converted to 1985 international dollars using 1985 purchasing power parities indexes (Summers and Heston, 1991).

had grown to more than half (53.6%) of public sector R&D spending worldwide. In 1991, Asian countries accounted for 62% of the developing world's publicly performed agricultural research expenditures (19% for China alone), the Latin America and Caribbean region and sub-Saharan Africa (including South Africa) each accounted for 12%, whereas West Asia and North Africa accounted for 14% of the expenditures.

Research intensities and spending ratios

An alternative perspective on agricultural R&D spending is provided by the agricultural research intensities (ARIs) presented in Table 11.2. The most commonly constructed ARIs express agricultural research expenditures as a percentage of agricultural GDP.[2] In 1991, as a group, industrial countries spent $2.39 on public agricultural R&D for $100 of agricultural output; a sizeable increase over the $1.47 they spent per $100, 20 years earlier. Developing countries, as a group, have much lower ARIs. In the early 1970s their ARI ratios averaged 38 cents per $100 of output, growing to only 51 cents by 1991. Other research intensity or spending ratios can be calculated: one measures agricultural R&D spending relative to the size of the economically active agricultural population, and another relative to total population. In 1991, industrial countries spent about $417 (1985 international dollars) per agricultural worker, nearly three times the corresponding 1971 figure. In 1991, developing countries spent $7.71 per agricultural worker, approximately twice the spending per agricultural worker in 1971. These differences are not too surprising, given the substantially higher proportion of developing country workers employed in agriculture and the more rapid contraction in the agricultural labour force in the industrial countries over the past several decades. Research spending per capita has risen too, by an average of 40% for industrial

Table 11.2. Agricultural research intensity ratios (%).

	1971–1975	1976–1980	1981–1985	1986–1990	1991
Developing countries	0.38	0.47	0.50	0.49	0.51
Sub-Saharan Africa	0.78	0.84	0.86	0.74	0.70
China	0.40	0.48	0.41	0.38	0.36
Asia and Pacific	0.26	0.36	0.44	0.50	0.55
Latin America	0.43	0.51	0.59	0.49	0.54
West Asia and North Africa	0.50	0.49	0.52	0.52	0.52
Industrial countries	1.38	1.60	1.98	2.18	2.39
Total[a]	0.67	0.76	0.81	0.79	0.81

Source: Pardey *et al.* (1997a).
[a] Excludes Cuba and former Soviet Union.

countries (from $6.30 per capita in 1971 to $8.84 in 1991), and by an average of 79% in developing countries (from $1.10 per capita in 1971 to $1.91 in 1991).[3]

Overview of global trends
Pardey *et al.* (1998a) summarized global trends, making the following key points:

- After a decade of strong growth, the annual growth in public investments in agricultural research slowed substantially during the 1980s (from 6.4% in 1971–1981 to 3.8% in 1981–1991 for developing countries, and from 2.7% to 1.7% for industrial countries).
- In 1991, developing countries as a group spent more ($8.0 billion) (1985 international dollars) than industrial countries ($6.9 billion) on public agricultural R&D, a reversal of the relative shares that prevailed only a decade earlier.
- The annual growth rate of investments in agricultural research by CGIAR slowed from 4.0% during the 1980s to 0.5% since 1990.
- The gap between agricultural research intensity ratios (i.e. public agricultural R&D relative to AgGDP) in industrial and developing countries widened considerably since 1971; by 1991 industrial country intensities averaged 2.39% compared with 0.51% for developing countries.

Industrial country trends

The dramatic shifts in public and private spending on agricultural R&D in industrial countries, and in the place of the agricultural sciences in overall spending on R&D, were summarized by Pardey *et al.* (1998a), with key points as follows:

- OECD countries as a group spent a total of $286.6 billion (1985 international dollars) on *all* public and private R&D in 1993; about two-thirds ($190.8 billion) was performed by the business sector. The substantial net flow of public funds to private research providers has diminished in recent years, in line with cutbacks in public funding for defence research.
- Public R&D spending priorities vary widely across countries. Notably, the share of central government R&D spending earmarked for defence research is down in most countries.
- The share of all public R&D funds going to agriculture shrank steadily from 8.9% in 1981 to 7.4% in 1993. Australia and New Zealand experienced a marked decline, whereas the rate of decline in the Netherlands, the UK and the USA was much more modest.

- Public plus private *agricultural* R&D spending in OECD countries totalled $14.1 billion (1985 international dollars) in 1993, up from $9.7 billion in 1981.
- Privately performed R&D is a prominent feature of contemporary agricultural research in rich countries. Private agricultural R&D totalled $7 billion (1985 international dollars) in OECD countries and accounted for about half of all agricultural research spending in 1993, and it grew at an annual average rate of 5.1% during 1981–1993.
- Private and public labs focus on different research. Less than 15% of public research is on food and kindred products, but 30–90% of private research has that orientation. Over 40% of private agricultural R&D in the USA and the UK focuses on chemicals and chemical products.
- Across the industrialized world, public spending on agricultural R&D grew by 2.7% per annum from 1971 to 1993 (slowing substantially to 1.8% per year from 1981 to 1993) to total $7.1 billion in 1993. Just 10 industrial countries (Australia, Canada, France, Germany, Italy, Japan, the Netherlands, Spain, the UK and the USA) now account for 90% of the industrialized world's public spending on agricultural research and 42% of global spending.

Economic Basis for Government Intervention in Research

The ideas presented in this section are developed in more detail in Alston and Pardey (1998). The net social benefits from research will be maximized when the marginal dollar spent on research yields a social benefit of $1. Many believe that from society's perspective, left to its own devices, the private sector would invest too little in agricultural R&D and would provide the wrong mixture of research investments. This is the market-failure argument for government involvement in agricultural R&D. To understand why agricultural R&D is a policy issue, and the relevant policy approaches, we need to understand the nature of the market failures that apply to knowledge and innovation. Market failure in research can arise from a variety of sources, which may be thought of in terms of: (i) divergences between private and social benefits and costs, caused by appropriability problems and externalities; (ii) economies of size, scale or scope in research; and (iii) the non-rival nature of new ideas and innovations produced by research.[4]

Divergent private and social costs and benefits

Market failure in R&D most often arises because private costs and benefits from research do not coincide with social costs and benefits. An important source of such divergences is knowledge spillovers or externalities, leading to appropriability problems, which means that some can free-ride on the R&D investments of others.[5] Such divergences cause *underinvestment*: R&D opportunities that would be socially profitable are not exploited because they are privately unprofitable.

Appropriability of research benefits
The main reason for private-sector underinvestment in R&D is inappropriability of research benefits: the firm responsible for developing a technology may not be able to capture (i.e. appropriate) all of the benefits accruing to the innovation, often because fully effective patenting or trade secrecy is not possible or because some research benefits (or costs) accrue to people other than those who use the results. Where research results are *non-excludable*, the innovator cannot fully prevent (exclude) others from taking advantage of the innovation. The practical problem is inadequate property rights; the related incentive problem is that it pays potential beneficiaries in the industry to wait in the hope of free-riding on the research of others, rather than doing or paying for the research themselves.[6]

Appropriability problems give rise to an asymmetry between the incidence of benefits and costs of research. For certain types of research, the rights to the results are fully and effectively protected by patents, so that the inventor, by using the results from the research or selling the rights to use them, can capture the benefits. Often, however, those who invest in R&D cannot capture all of the benefits; others can free-ride on an investment in research, using the results and sharing in the benefits without sharing in the costs.[7] Hence, private benefits to an investor (or group of investors) are less than the social benefits of the investment and some socially profitable investment opportunities remain unexploited. The upshot is that, in the absence of government intervention, investment in agricultural research is likely to be too little. Complete appropriability of benefits may give rise to a different set of problems, related to the exercise of monopoly power in new technology.

It is not a question of absolutes. Complete appropriability of the fruits of invention is unlikely ever to be possible. So long as some benefits can be obtained there will be incentives for some investment. The real issue, then, is the appropriate form and degree of partial correction for problems arising from incomplete appropriability. Since intervention is justified only when the benefits exceed the costs, in many cases the costs will be such that no intervention is the best policy.

INTERNATIONAL SPILLOVERS The appropriability problem extends beyond relationships among individuals within a state or nation. The knowledge and technology derived from agricultural R&D does not respect geopolitical boundaries, and technology spillovers among states of a nation or among nations give rise to incentive problems in research. For instance, several studies (e.g. Brennan and Fox, 1995; Pardey *et al.*, 1996, 1997a) have documented evidence that benefits from wheat and rice research conducted in international centres have yielded major pay-offs in Australia and the USA as well as in developing countries. These studies have shown huge returns to this investment, from both a global perspective and from the perspective of donors who have been 'doing well by doing good' (Tribe, 1991).[8]

INTERINDUSTRY SPILLOVERS The appropriability problem may take another form: spillovers through interindustry applications of research results, such as when livestock R&D has applications for human medicine or vice versa or when fundamental findings related to biosciences are applicable across a broad range of commodity research. Again, the problem is that these spillovers usually will not be recognized as benefits by the providers or funders of the R&D. Essentially, this is an interindustry version of the inappropriability problem described above.

Distortions in other markets
Divergent private and social benefits from research may also arise from a more general set of distortions in the economy, owing to distortions caused by government policies (such as agricultural commodity programmes or trade policies) or caused by a lack of government action to correct a distortion, such as those arising from market power of firms or from environmental externalities.

ENVIRONMENTAL EXTERNALITIES Spillovers and externalities may be relevant not only in relation to the products from research – a direct distortion in the research enterprise – but also through distortions in the markets for agricultural outputs and inputs leading to indirect distortions in research. Agriculture often involves environmental externalities arising from spillover effects of agricultural production on other agricultural producers (e.g. through effects on incidence of pests) or others through impacts on groundwater, air pollution or food contamination which are not compensated for through markets. Even in the absence of market failures associated with the atomistic nature of agricultural production and appropriability, there will be distortions in incentives so that the direction of research will be biased against environmental externality-mitigating technologies and in favour of environmental externality-exacerbating technologies. Hence, in the absence of government intervention, commercial decisions may produce too much pollution and preserve too little wilderness. There is too little R&D due to inappropriability and the mixture of R&D is biased because of externality effects (for an analysis, see Alston *et al.*, 1995a).

If the adoption of new innovations or techniques exacerbates environmental problems, then some might say that there is too much rather than too little agricultural R&D. However, as noted by Lloyd *et al.* (1990), Alston *et al.* (1995b) and Alston and Pardey (1996) this is not the correct inference. Rather, these studies suggest that the 'mix' of research is suboptimal; in particular, there is likely to be a case for research into those environmental problems causing concern.[9] But it is important not to overlook the fundamental source of the problem, the externality, and not to blame research for exacerbating the costs of the externality; it is the failure of government policy to correct the externality which is the real problem.

Agricultural R&D can generate technologies that are environmentally friendly, relative to the current technology; but it is not sufficient to invent the technology. The very nature of (negative) externalities means that private investors do not have adequate incentives to make an effort to reduce them, either through the choice of production practices with available technology or through the choice of the direction for technology to evolve through research, development and adoption decisions. If agricultural R&D is to be effective in reducing environmental externalities, the resulting new technologies must be adopted, and if they are to be adopted, they must be viewed as privately profitable. This could happen in one of two ways: a new (environmentally friendly) technology may be privately more profitable than the current technology under the current incentives, or the government may act to change the adoption incentives as well. Similar arguments apply to the development and adoption of technologies that consume stocks of unpriced or underpriced natural resources. Private incentives are liable to lead in the direction of the development and adoption of excessively consuming technologies unless government acts to modify the incentives and 'internalize' the externalities.

COMMODITY POLICIES A substantial literature now exists concerning the effects of commodity price policies on the size and distribution of research benefits and the corresponding incentives for private and public investments in different types of research. Much of this literature is reviewed and summarized in Alston *et al.* (1995b). The central result is that the main effect of most types of price-distorting policies is on the distribution of benefits and costs of research (i.e. who benefits) with relatively small effects on the total benefits; social benefits from research are less, in the presence of a price-distorting policy, by an amount equal to the effect of the research in question on the social costs of the policy. Like environmental externalities, it is tempting to blame research when its adoption leads to a worsening of the social costs of given market distortions arising from government price-support programmes; but the real culprit is the price support programmes.[10]

The existence of government policies that distort prices may change the calculation of benefits and costs of research, and change the incentives of different groups in society to conduct R&D or adopt the results. There are no clear general implications, however, for whether the tendency to over- or underinvest in research is greater or smaller, or whether the social pay-off from public research investments is greater or smaller as a result of commodity price support policies. This counters the common view that positive protection to producers of a commodity means that public investment in research on the commodity should be less than otherwise.

MARKET POWER OF FIRMS A closely related issue is the effect of commodity–monopsony power of processing firms buying from farmers, or monopoly power of firms selling inputs to farmers, or selling processed agricultural products to consumers. Market power of firms can lead to a divergence between private social costs of the commodities affected and, accordingly, in the social and private benefits and costs of agricultural R&D. There is some evidence of market power of firms in some agricultural industries, but there has been relatively little work undertaken to establish its effects on returns to agricultural R&D.

Recent work by Alston *et al.* (1997) has shown that distortions arising from market power of firms affect the returns to research in much the same way as distortions arising from commodity policies: the main effects are on the distribution of benefits; the effects on net social benefits from R&D are equal to the effects of the R&D on net social costs of the distortion; and there are no general rules since the nature of the research-induced change in technology, functional forms of supply and demand, and other details of the commodity market, all matter. Market power of firms may have more important consequences through its implications for the likelihood that firms will be able to appropriate the returns from invention. That is, as the food processing sector, for instance, becomes more concentrated, it is more likely that individual firms will find it profitable to invest in technology that can be only partially protected by patents and trade secrets. Thus, while it may lead to greater distortions in the commodity market, with mixed implications for the total investment in R&D, increased concentration in food processing is likely to lead to a greater private investment in R&D and a reduction in the distortion that arises from inappropriability and consequent underinvestment in food processing research.

Industry structure and economics of agricultural R&D

Some incentive problems arise from the economics of the research enterprise as it relates to the size of firms. The nature of research activity, which is usually long term, large scale and risky, means that the typical firm in agriculture is not able to carry out effective research (although it can help to fund it), and institutions may need to be set up on a collective basis.[11] This is a particular problem when economies of size, scale and scope in R&D are significant, but individual farmers cannot exploit them and transaction costs prevent farmers from acting collectively. There can be advantages to centralizing R&D resources and infrastructure, which firms in concentrated industries may be able to exploit; but this centralization is unlikely to be achieved by competitive producers in atomistic rural industries. The reason is that, without statutory backing, the transactions costs of coordinating and enforcing the collective agreements needed to exploit the relevant economies would probably be prohibitive.

Economies of size, scale and scope
Economies of size, scale and scope may be greater in agricultural R&D than in other industrial R&D because of the biological base of the industry, long production cycles and jointness between agricultural enterprises.[12] Whether economies of size, scale or scope are greater in agricultural R&D than in other industrial R&D, or the industry structure means that transaction costs are larger, or for both these reasons, market failure from this source is likely to be relatively important in agricultural R&D.

R&D beyond the farm gate
Farm input suppliers and other components of the agribusiness industry that transport, process and market farm products, tend to be relatively large firms which are more able to exploit economies of size, scale and scope in R&D. Agribusiness technologies tend to be mechanical, of the type that can be protected by patents, or process innovations that can be protected by secrecy. The technology used by agribusiness is often not specialized to agribusiness and can be adapted from broader industrial technologies (e.g. refrigeration or transportation technology). Thus, the potential role for the government (i.e. the odds of market failure) is *generally* greater in R&D pertaining to

farming than in R&D pertaining to agribusiness.[13] There are exceptions. Some parts of the farming industry are involved in vertically
integrated structures where research benefits can be internalized (e.g.
the broiler chicken industry); certain types of technology applicable
to farming are effectively protected by patents (e.g. machinery). R&D
incentive problems are important in some parts of agribusiness (e.g. in
plant breeding there is 'natural' appropriability for hybrid lines, since
the crop does not reproduce itself, but for open-pollinated varieties
property rights must be legislated and enforced to ensure appropriability). It is important to exercise discretion in judging where the
market failures in R&D are important and where they are not, since
government investment in R&D in a particular area is likely to crowd
out some private sector R&D. Where private sector underinvestment in
R&D is not otherwise a problem, public sector R&D can *cause* private
sector underinvestment.

Non-rival goods

The final important source of market failure in research is that knowledge, the product of research, often takes a form that makes it *non-
rival* in consumption: the use of the knowledge by one person does not
diminish its value to another. Non-excludability occurs when property
rights in a good have not been, or cannot be, defined. Non-rivalry
exists when one person's consumption of a good does not detract (or
subtract) from someone else's consumption of it. Some goods, including some types of research output, may exhibit some degree of non-
rivalry or non-excludability (and thus they will be partially rival or
excludable). These properties should not be thought of in absolute
terms. The market will tend to overprice non-rival (or 'public') goods
from society's perspective.

A potentially difficult trade-off must be faced.[14] The difficulty
arises when we attempt to remedy the non-excludability problem
directly by strengthening the property rights applying to research
results (e.g. by patents or copyrights), thus encouraging greater private
provision of R&D by enabling research producers to reap the rewards
from their efforts. Even if it had no practical difficulties, this approach
often clashes with the most appropriate arrangement for provision of a
non-rival good, which involves its provision at zero price since the
social marginal cost of providing it is zero.[15] In other words, the price
that is high enough to encourage innovation is (definitionally) too high
from the point of maximizing the social benefits of the innovation.[16]
Public providers attempting to raise revenue often appear to ignore
this principle in setting prices for information that has public
good characteristics. For these reasons, 'privatization', in the sense of

making research results excludable by improving the definition of property rights attached to those results, can involve losses in economic efficiency because it does not address the other characteristic of public goods: non-rivalry in consumption.

Evidence of distortions in total investment and the mix

Much can be learned about the likelihood of market failure in research from examination of the conditions in the relevant industry, the nature of rights to invention and the institutional structure, in the light of the discussion above. But this constitutes no more than a first step in establishing the *importance* of market failure in research and the desirability of government action to reduce the social costs of market failure in research. The harder, remaining questions are empirical ones: what would be the net social benefits from various alternative policies? Some work by economists has provided partial answers to some of these questions.

Several studies have surveyed the evidence from benefit–cost studies that have reported rates of return to agricultural R&D (e.g. Echeverría, 1990; Huffman and Evenson, 1993; Alston and Pardey, 1996). The conclusion from these studies is that the rates of return to public investments in agricultural R&D have been high; high enough to justify an even larger public investment in research. The same evidence also establishes that the in-principle arguments suggesting that the private sector would underinvest in agricultural R&D are borne out: even with the substantial public investments, the rates of return have been high enough to justify even greater investments.

Unfortunately, the evidence from the past does not provide much more guidance than this. For instance, many interpret the evidence as justifying a greater investment in research. But we cannot say from this evidence whether the public investment should increase by 10%, 100% or even more (indeed, some would say that since the evidence is historical, it cannot even be used to say that public research investments necessarily should increase now or in the future). We have no real information on whether the rate of return to research is diminishing with respect to the total annual investment or diminishing (or increasing) over time as a result of the accumulation of knowledge. Other empirical and essentially unanswered questions concern the social rates of return to public versus private research, to basic versus applied research, to research versus extension, to agricultural research versus more general research, to livestock versus plant research and to environmentally oriented versus production research. We have some evidence supporting our prior expectations, that relatively high rates of return accrue to more basic (and less appropriable) research, and that public research tends to earn a higher social rate of return than private research does. But realistically speaking the evidence is weak on all these comparisons. There is no evidence of the rate of return to environmental research justifying a shift of resources in that direction, for instance.[17]

Some of the measurement problems are essentially intractable given currently available data. Hence, in some cases, decisions about the extent of market failure and appropriate intervention must be based on informed judgement alone. In some other cases, however, reasonably reliable estimates of likely rates of return to investments can be calculated. Hence there is likely to be a significant social pay-off from further investment in developing the methods and data required, and the estimates themselves, to support research policy decisions.

Principles for Intervention

General principles

The optimal intervention by a government whose aim is to reduce the distortions arising from inadequate private-sector incentives for agricultural R&D, would:

- optimize the total investment in public-sector agricultural R&D and the mix of R&D, while minimizing the distortions from crowding out private R&D;
- minimize the costs of raising revenues to finance public-sector R&D by using least-cost sources of funds;
- organize public-sector R&D institutions so that they can conduct R&D in the least-cost way, according to their comparative advantages, with a minimum of wasteful replication of facilities and programmes, but taking account of synergism between public-sector R&D and other public-sector activities such as higher education;
- allocate and use research resources efficiently among programmes and projects (i.e. according to economic, not political criteria), minimize transactions costs and administrative and bureaucratic overhead, and allow decentralized decision making where effective incentive mechanisms are possible.

These four principles relate to economic efficiency of R&D in terms of: (i) the total funding; (ii) the sources of funds; (iii) institutional organization; and (iv) resource allocation and management. Some more specific points outlined below, relate to particular aspects of the general agricultural R&D policy issues.

Specific points

Multiple distortions

First, in devising policies to reduce underinvestment in R&D, it may be necessary to pay attention to several types of market distortions, including those related directly to R&D arising from incomplete appropriability of research benefits, those arising from pricing of non-rival goods produced by R&D above their marginal social cost, and the more general set of distortions in the economy arising from government policy and environmental externalities.

Separate financing and research roles

Second, a distinction between the ideal arrangements for *provision* (production) of research and the ideal arrangements for *financing* research must be kept in mind when designing remedies for a private-sector underinvestment in R&D. Certainly, doing the research and paying for it are related and should be considered in conjunction, but it is worth keeping the distinction in mind, for the following reason. Research funds should be raised in a least-cost manner; that is, the sum of administrative costs and distortionary losses to the economy should be minimized. The issue of how to spend the funds is clearly separate from how they should be raised. Raising the funds in the least-cost way should not preclude the use of efficient competitive processes for allocating them among research programmes.

Beneficiary pays

Third, to the extent practicable, the *beneficiaries* of research should bear the cost. On this principle, when most of the benefits from a research programme accrue within an industry, we would expect the industry to finance the research; when most of the benefits spill over to the general community, we would expect the research to be publicly financed. There are two reasons for employing the *beneficiary-pays* principle. One is based simply on a notion of fairness; that if beneficiaries can be identified, they should foot the bill where that can be facilitated. Further, it may be cheaper (more efficient) to raise money from the relevant industry for R&D, rather than using general public funds raised through the tax system (Alston and Mullen, 1992).[18] Thus, in plausible circumstances, there are equity *and* efficiency arguments in favour of the beneficiary-pays principle. The economic efficiency arguments are paramount.

Economic way of thinking

Economic efficiency also requires allocative efficiency for a given set of research resources. This means allocating research resources to the activities with the highest expected pay-off and providing a set of management processes and incentives such that the resources are not wasted. Ultimately, as argued by Alston *et al.* (1995b), the greatest gains are likely to come from inculcating an 'economic way of thinking' into research resource allocation processes at all levels, rather than from formalized bureaucratic processes for planning and accountability; although formalized processes must also play a role.

Do no harm

Finally, an overriding principle is that any remedies for market failures should clearly make matters better in aggregate, in the sense that the costs associated with the remedies should be outweighed by the benefits. With this in mind and in recognition of information problems and potential for 'government failure' as well as market failure, interventions should be designed to maintain or increase competition in the provision of research services, and transparency in the processes through which public R&D resources are allocated.

Guidelines for intervention

The last point also reminds us that no action is often the best policy, even if we are disappointed in the outcomes of the unfettered workings of the free-market mechanism. It may be that we wish that the private sector would invest more in certain types of agricultural R&D, but that the alternative of doing nothing is better than any feasible interventions that can be designed. It is one thing to establish a case of market failure. It is another to determine the best action for the government to take, indeed, taking *no* action may be the optimal policy. An intervention is justified only if it improves the situation by reducing social costs of market failure: the benefits of the intervention must be greater than the costs. This perspective was illustrated in the State Government of Victoria (1994) submission to the Industry Commission inquiry into R&D in Australia, which provided a decision tree (Fig. 11.1) that can be used for deciding whether to intervene in agricultural R&D and how.

The first step is to establish a *prima facie* case of market failure. If the market is performing its functions adequately, questions still may arise about whether the market outcomes are satisfactory. If the market is working and the outcomes are acceptable, then the government should not be involved. Even if the market is not working ideally, or the outcomes are not totally satisfactory, the case for government involvement depends on whether the problems with the unregulated situation are serious, and whether the perceived problems are likely to be amenable to correction by policy. In many cases the answer at this stage would be to accept the existing market imperfections rather than

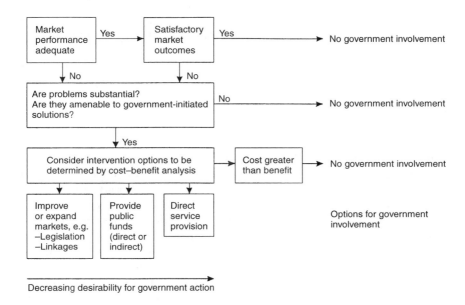

Fig. 11.1. A decision tree for government intervention in cases of market failure. Source: State Government of Victoria, Australia (1994, p. 14).

make things worse by intervening. In some cases it is worth proceeding to the next stage of decision making, and to consider explicit policy options. Even at this last step in the decision-making hierarchy it might be concluded that the costs do not justify the benefits from intervening. In some cases, however, intervention is justified by the benefit–cost assessment. Agricultural R&D provides a number of examples.

Intervention Options

Many interventions are possible and, in fact, are used in relation to agricultural R&D. The options available for government action to mitigate underinvestment involve some combination of the following:

- Undertake the R&D itself using general government funds.
- Provide funds or other financial incentives (subsidies, tax credits) for R&D.
- Create institutions or arrangements that facilitate private funding in order to skirt the problems of transactions costs and free riding

without overpricing non-rival goods (e.g. facilitating the development of marketing-orders or cooperatives and appropriate institutions for administering the funds).
- Enhance property rights to improve the private incentives for conducting R&D (e.g. patents and plant variety rights).

Taxpayer funding of R&D

A mix of private and public funding is used to support national and international agricultural research and in many countries national and provincial governments conduct separately administered programmes of research. The primary source of funding for these expenditures is general tax revenues. Decisions must be made about how much to spend on agricultural R&D from general revenues of the government and how to allocate those resources among alternative agricultural R&D investments. Such decisions cannot be made in isolation from other elements of the agricultural R&D policy, such as the mechanisms in place to encourage private sector investments, the institutional setting and property rights.

Optimal agricultural research intensities

Some seek simple solutions, such as a rule of thumb that agricultural research intensity (ARI) ratios, which measure agricultural research spending relative to the value of agricultural output, should be at least 2%. Such a rule would imply significant increases in ARIs in most developing countries and such an outcome may well be appropriate, but this does not mean the 2% rule is right. Unfortunately the world is not so simple. Should all ARI ratios be equal everywhere? This is the congruence rule which can give some guidance to the incongruence of existing research resource allocations, as a basis for reconsidering those allocations based on evidence about relative rates of return, but should never be seen as an objective criterion for an efficient total quantity of research funding or a rule for efficient allocation of research resources between research areas. A second level of decisions concerns the allocation of the total research resource between alternative uses: different institutions, programmes or projects, with different scientific, commodity or institutional orientation. Getting the allocative efficiency right may be even more important than optimizing the total quantity of research resources.

While it is easy to criticize congruence and other rules of thumb, it is not easy to come up with clear guidance about how to determine how much should be spent from general government revenues on agricultural R&D in total or in particular areas of research. The rule to have in mind is that it is appropriate to spend as much as can be spent profitably, that is, keep spending so long as the social rate of return exceeds the social opportunity cost of the funds, taking care to consider the implications of public spending for crowding out private investments and so on. The criterion, then, is the expected social rate of return and how to assess that will depend on the circumstances (e.g. see Alston *et al.*, 1995b), but some subjective judgement is inescapable.

Forms for funding

A related question concerns the forms in which taxpayer funds are distributed to research providers. Unfortunately, there is little hard evidence on the consequences of the implications of different processes and procedures for allocating research resources on the net benefits. The possibilities take the following general forms:

- 'Gifts', or 'formula funds' which are funds provided with no particular strings attached.
- 'Grants', which specify some general commitments by the researchers.
- 'Contracts', which entail specific obligations.

There is a perception that recent years have seen moves towards proportionately greater use of contracts and grants and a reduction of gifts (i.e. formula funding). Has the desirable balance been overshot?

COMPETITIVE GRANTS Competitive grants have a great deal to recommend them as a way of allocating public sector research resources. However, competing for grants is hard work and expensive, and if competitive grants are to deliver the promised benefits of greater allocative efficiency, they have to be allocated according to efficiency criteria. A poorly administered and corrupt system of competitive grants could easily be worse than an inflexible system of block grants or funding according to some formula, unrelated to past or prospective performance. Some (e.g. Just and Huffman, 1992; Huffman and Just, 1994) have argued that formula funds, which are essentially gifts in that the funds are provided to scientists without regard to their performance, are more efficient than competitive grants. Their primary concern is that transactions costs involved in competitive grants programmes – in terms of the costs to individual scientists of preparing proposals, and reporting to granting bodies, and the costs of evaluating the proposals and deciding which ones to support – are so high that the programmes cannot be economic. That charge could be correct; but relevant alternatives must be compared, and on a comparable footing.[19]

Every method of allocating research resources involves four types of costs: (i) information costs (the costs of obtaining relevant information on the benefits from different types of R&D projects, on which to base decisions); (ii) other transactions costs (the costs of applying for grants, managing them and administering them); (iii) opportunity costs of inefficient resource allocation, due to research resources not being used in the projects and programmes with the highest social pay-off; and (iv) rent-seeking costs (costs of resources being spent waste-fully attempting to cause a redistribution of grant resources). Different research resource allocation processes will involve different amounts of particular types of costs. For instance, through the proposal process, competitive grants generate information about research alternatives for decision-makers. Although they may lower the cost of certain types of information, they also involve relatively high transactions costs. They might also involve relatively high rent-seeking costs (for instance, scientists lobby for support). However, these additional costs may be justified if competitive grants lead to a lower overall social cost because they reduce the (opportunity) cost of resource misallocation. On the other hand, formula funds involve relatively high resource misallocation costs, which get higher the longer a formula stays fixed (since circumstances change), and relatively low transactions costs. This is not to say the transactions costs are zero, or that rent-seeking costs are zero with formula funds (there is a fair bit of bureaucracy associated with the administration of the funds; the formulae do or, at least, may change from time to time; some resources are spent simply to preserve the status quo). Earmarked funds may involve the greatest rent-seeking and resource distortion costs, but they may also involve relatively small transactions costs. In short, the full costs should be considered when comparing research resource allocation procedures.

PEOPLE VERSUS PROJECTS A middle ground is likely to be best for many situations: enough competition to ensure a vigorous and adaptable research programme, which exploits optimally the available informa-tion on scientific opportunity and economic implications; enough security and confidence in future funding so the scientists will take appropriate risks, pursuing long-term opportunities; not too much cost of time of scientists involved in drafting proposals, justifying expen-ditures and reporting results; not so narrow-minded that curiosity and flair are stifled.

Such an optimal situation, with every element just right, is hard to achieve. Part of the solution is likely to involve relatively long-term funding of particular people, rather than particular projects, based on their past performance more than their promises about the future, per-haps especially for the more basic types of scientific work. Competi-tion can be effective as a resource allocation and incentive mechanism without requiring a morass of planning processes and committees, which to some represent the antithesis of competition.

Alternative institutional arrangements for funding R&D

Recent studies (e.g. in the USA and Australia) have shown that it costs society well over a dollar to provide a dollar of general taxpayer revenues to finance public expenditures; in less industrial countries, the taxation systems are likely to be less efficient and the marginal opportunity cost of government spending is likely to be even higher than in the industrial countries.[20] It is also an increasingly scarce source of revenues. Alternative sources of revenue may be less expensive, fairer and politically more sustainable, when used to finance certain types of research and to achieve an expanded total public-sector R&D budget.

Different agricultural R&D programmes and projects call for different funding arrangements. Agricultural R&D may be a public good in the sense of (at least partial) non-excludability and non-rivalness, but this does not mean that everybody in a nation, or the world, benefits and it does not mean that everybody should pay. Both fairness and efficiency are promoted by funding research so that, as much as possible, the costs are borne in proportion to the benefits. This can be promoted by choosing funding arrangements that reflect the geographic focus and the commodity orientation of the research, which implies a greater use of subnational and multinational, regional or commodity R&D programmes.

Check-off funding

Alston and Pardey (1996) argue that a greater use of commodity check-off funding is warranted for three reasons. First, industry funding is a potential complement to other sources of funds which, as a practical matter, are likely to continue to leave total funding inadequate. Second, commodity check-offs are likely to be a relatively efficient (and fair) tax base.[21] Third, industry funding arrangements can be organized to provide incentives for efficient use of both the check-off funds and other research resources. Check-off funding is clearly applicable to research on a particular commodity. By definition, this is not basic research. Check-offs are clearly more applicable for commodities that are traded through markets in developed industries and less applicable to, say, subsistence crops or commodities for which markets are not well developed and where, in consequence, the costs of raising check-off funds would be prohibitively expensive. Similarly, check-off schemes tend to be less applicable to research that affects multiple commodities and research that applies to particular factors of production or that has an environmental focus. However, these issues notwithstanding, commodity check-offs could be used more extensively to support the significant proportion of research that can be identified with a well-defined commodity (or other) interest group. Around the world, such mechanisms are relatively underutilized in the sense that only a small fraction of total R&D resources are generated in this fashion.

Mixed private–public funding mechanisms

A number of options exist whereby the government can devise policies that promote a mixture of private and public funding for research, and this can be an effective mechanism for leveraging a greater total research effort from a given set of taxpayer-based research resources, and for assuring an equitable distribution of costs when research benefits are partially appropriable within industry.

MATCHING GRANTS Matching schemes, whereby funds from selected sources are matched on some basis with public funding to support agricultural R&D have been practised for some time. The Bankhead Jones Act of 1935 established that US federal government funding earmarked for the state agricultural experiment stations would require matching state government support. Since 1985 the Australian government provides federal government funds to Research and Development Corporations (RDCs) on a dollar-for-dollar matching basis (up to a ceiling of 0.5% of the gross value of production of the respective industry served by the RDC) against producer levy funding. The idea behind these matching grant schemes is that the effects of agricultural R&D spill beyond the specific state or industry actually carrying out or providing support for the research. So, in the absence of matching federal government support to address these appropriability problems too little research would be performed and the nature of the R&D may be more narrowly focused than would be socially optimal. Difficulties in designing a matching grant scheme involve deciding on the rate of support (i.e. the appropriate basis by which to provide matching funds – dollar for dollar, one for two or something else), the ceiling level (if any) of public support to provide and the appropriate public role in the allocation of the pooled public and private funds (Industry Commission, 1995, Vol. 2).

COST-RECOVERY AND CONTRACT RESEARCH Another way to finance public-sector agricultural R&D is to sell scientist time or the results of research. To the extent that exclusion is possible (i.e. where the knowledge that results from research is not a _pure_ public good, and property rights are enforceable), it is possible to collect royalties for the use of research results. It is usual for public sector organizations such as universities or government research institutes to patent their research results where possible, and to license the technology. Many universities (and other public agencies as well) have set up their own technology transfer offices on campus, or hire outside expertise to deal with the management of intellectual property on their behalf. Is this a good idea? A number of reservations and caveats should be raised.

The general principle of efficient pricing is that the government should charge the marginal social cost for the goods it produces. Thus, when the outcome of research is a pure public good, non-rival in the sense that one person's use does not diminish its availability to others (such as televised weather forecasts), the marginal cost of supplying it to another individual is zero and it should be provided gratis. In other cases it may cost something to make the good available (e.g. costs of multiplying seed) and it is appropriate to recover those costs, but it would be inefficient to charge as much as the market would bear. Complications arise when research is undertaken as a joint venture between government and industry. Other arguments concern 'second-best' corrections for problems of inadequate scale or scope – some commercial activities help to spread the overhead cost of lumpy capital (such as research facilities, equipment, or even scientists) in a public-sector institution with unused capacity – or inadequate funding due to 'government failure'.[22]

One option is that public-sector R&D institutions should avoid investments that are demonstrably profitable in the sense that costs can be recovered from the sale of the results. In such cases, the private sector should have adequate incentives and government intervention will either crowd out private investment or, worse, be directed into an area that is not profitable, either privately or socially. The in-between cases are most troublesome; cases where there is potential for some cost recovery but where there are grounds for believing the private-sector investment will be inadequate.

TAX CONCESSIONS The provision of tax concessions for private research, which has been a common policy in a number of countries, is another form of joint-venture public and private funding of research. It is a blunt instrument, in that it is difficult to discriminate closely among alternative forms of research, but it is relatively inexpensive to administer since, in most places, the tax system is well-equipped for such purposes.

Public roles regarding the performance of R&D

Alston *et al.* (1998b) note that public agricultural R&D has undergone various degrees of restructuring in recent years in the five OECD countries for which detailed reviews have recently been completed.[23] The Netherlands, New Zealand and the UK have substantially revamped the public agencies involved in carrying out research. In England and Wales, as well as in the Netherlands, the agencies that perform public R&D have been consolidated and, in some important instances, commercialized. In New Zealand a number of new Crown Institutes have been established in an effort to develop a public 'market' for R&D services that distinguishes between buyers and sellers of such services. The public agencies that undertake agricultural R&D in Australia and America have not changed as much, although the funding and management structures within Australia have been radically revised in recent years (see Clarke, Chapter 4, this volume; Henzell et al., Chapter 7, this volume).

Conclusion

Over the past few decades OECD countries have been experimenting with new ways to provide public support for agricultural R&D and reshaping the public roles in performing that research. Not all aspects of these changes are likely to be socially beneficial for the countries concerned nor are all these aspects directly relevant for other (especially developing) countries looking to revise their public policies regarding agricultural R&D. Unfortunately, the long lag times involved in doing agricultural research, having the results adopted and realizing the returns means that it is difficult to judge the pay-offs to these new policies and practices. Nonetheless, the forces that gave rise to many of these changes are being experienced by many countries, albeit in differing degrees and with different domestic circumstances that would shape the appropriate policy response.

The policy options outlined in this chapter are intended to guide policy-makers in responding appropriately to the changing circumstances surrounding agricultural R&D and the public role in it. Many of the issues are subtle and complex. They go well beyond considerations regarding the size of the respective public and private commitments to funding agricultural R&D. Major issues concern the form in which the funds are raised and allocated, the public versus private roles in performing (as distinct from financing) the research, and the links between R&D and other government policies that affect agriculture. Changes in the science of agricultural research, the property protection afforded to new technologies and the global market structures regarding agricultural input and post-farm firms also have a bearing on these public policy decisions. Thus all these decisions involve institutional and incentive effects that can have significant short-term and potentially profound long-term consequences on the performance of the agricultural sectors that directly draw on the R&D. Ideally, proposed policy options should be assessed carefully, monitored periodically and revised when evidence suggests they are not working as envisaged or that alternative, superior policies are feasible.

The principles laid out here provide a framework for this process of ongoing policy review.

Notes

[1] These 'global' totals are preliminary estimates that exclude Eastern European and former Soviet Union countries. The principal data source for the 1961–1985 period is Pardey *et al.* (1991). These data were revised and updated for African countries using various ISNAR *Statistical Briefs*, for most of the principal Asian countries (including China and India) with data from Pardey *et al.* (1996), and for the industrial countries, from Pardey *et al.* (1998a). Semiprocessed data from numerous other sources were obtained for most of the mid- to larger-sized NARS and a number of smaller systems. The developing countries for which we have direct estimates account for approximately 85% of the developing country total.

[2] Agricultural GDP is a 'value added' measure of agricultural output which represents the gross value of output minus the value of purchased inputs such as fertilizer, pesticides and machinery. Thus these research intensities are higher than, and not directly comparable with, other research intensities that divide agricultural research spending by the *gross* value of output.

[3] Alston and Pardey (1993) report comparable measures that they use to discuss the political economy aspects of public investment in agricultural R&D.

[4] It is not sufficient to find a departure from the conditions of perfect markets to claim that markets have failed. No real-world market can stand up to the comparison against the textbook perfect market. Rather, one must compare the 'imperfect' market outcome with a relevant real-world alternative, taking into account the information problems and transactions costs that are part of the real world.

[5] Externalities arise when one individual's production or consumption activities involve *spillover* effects on other individuals that are not compensated through markets. Groundwater pollution with agricultural chemicals is an example of a *negative* externality. Free-riding by others on an individual's research results is a type of externality too, a *positive* externality which has beneficial spillover effects. The existence of externalities means that *private* costs (or benefits) from economic activities differ from the corresponding *social* costs (or benefits) and, as a result, private decisions are not socially optimal: a market failure.

[6] Such problems extend beyond the laboratory to adoption processes that involve trial and error in farmers' fields. For example, Foster and Rosenzweig (1995) found that farmers in India learned about new high-yielding (green revolution) varieties of wheat and rice both from their own experience and the experience of their neighbours. This meant that the farmers had an incentive to delay their own adoption of varieties and free ride on their neighbours' investments in learning about the new technology, a market failure leading to a delay in the adoption process.

[7] For instance, the benefits from most mechanical inventions and developing new hybrid plant varieties, such as hybrid corn, are mostly appropriable. On the other hand, an agronomist or farmer who developed an improved wheat variety would have difficulty appropriating the benefits. The inventor could

not get the potential social benefits simply by using the new variety himself/ herself; but if he/she sold the (fertile) seed in one year the buyers could keep some of the grain produced from that seed to use as seed thereafter. Venner (1997) documents these arguments in detail and reports sustained inappropriability problems in US wheat varieties, even under the Plant Variety Protection Act.

[8] For instance, Pardey *et al.* (1997b) showed that by the early 1990s, about one-fifth of the total wheat acreage was sown to varieties with CIMMYT ancestry and around 73% of the total US rice acreage was sown to varieties with IRRI ancestry. This meant, for example, that US wheat producers gained at least $3.4 billion over 1970–1993 from CIMMYT wheat variety improvements, which implies a ratio of benefits to costs borne by the USA of at least 40:1. The same study found that California alone could have profitably financed the entire CIMMYT research programme, given the benefits flowing to California's wheat producers.

[9] These ideas have been elaborated upon more recently by Ervin and Schmitz (1996), although they go somewhat further in proposing that the distortionary effects are so great that there is a negative rate of return to research, and in suggesting that the share of research devoted to environmental issues has been too small relative to commodity-specific research. We challenge their data (which considers only USDA funding of natural resource and environmental research – see Alston *et al.* (1997) for a more complete analysis) and their conclusion (there is no evidence to support the view that the social rate of return would be raised by redirecting the research resources in the direction they suggest).

[10] If one blames R&D for worsening the social costs of price support programmes, who takes the credit when the government eliminates a programme and the benefits from doing so are so much greater than they would have been if technology had never changed?

[11] There are exceptions to the *typical* situation, but even when firms are large enough to find it profitable to carry out some research there is still likely to be too little research for the other reasons (appropriability and externalities).

[12] Economies of scale are exploited when the research organization is large enough to be able to specialize and allow appropriate division of labour. Economies of size are exploited when the R&D enterprise is large enough to have appropriate combinations of lumpy fixed factors, such as physical infrastructure and scientific expertise, and other variable inputs, such as equipment and support staff. Economies of scope refers to benefits from diversification of the R&D portfolio. Scientists in different fields may be able to make contributions to the research of others; or items of equipment or infrastructure may be able to be used for different projects at different times of day or night, or in different seasons of the year.

[13] Some proponents of value-added production have proposed other forms of government support for it, not just R&D. Cashin (1988) criticized this more general value-added movement from a more general comparative advantage perspective. His arguments are persuasive.

[14] In the case where the benefits are industry-specific, but inappropriable to private R&D providers, the situation may be dealt with simply as if transactions costs are the problem; that is, facilitate private (industry-level) funding

and allow the results to be disseminated freely, for reasons discussed below. This enables the costs and benefits of R&D to be internalized within the industry in a fairly straightforward manner.

[15] A standard result in economics is that efficiency in a market is achieved when the price of the good is set equal to the marginal cost of providing it. Thus when a good can be provided to further people at no extra cost – it is completely non-rival in consumption – efficiency dictates that it be provided at a zero price.

[16] This is not to mention the problem of potential overinvestment in R&D, as firms compete with each other to win the patent first. The idea of 'patent races' is discussed in Wright (1983).

[17] A current project by Alston *et al.* (1998a) is documenting a more comprehensive review and analysis of the evidence on rates of return to agricultural R&D.

[18] As Alston and Pardey (1996, pp. 254–255) show, the marginal and average excess burden of taxation rises with increases in the supply and demand elasticities and the tax rate; the total excess burden rises with the square of the tax rate. It follows that, in the absence of any other distortions, a small tax on an agricultural commodity must have a very small total, average and marginal excess burden compared with general taxation measures.

[19] For instance, even though economists often disagree on these general questions they are mostly united in their condemnation of earmarked funds. The implicit assumption is that the earmarked funds could be reallocated to some research area with higher social pay-off; but perhaps if they were not earmarked funds, politicians would be less interested in providing them. Even earmarked grants, tainted by political processes, may be better than no grants at all.

[20] Fullerton (1991) and Ballard and Fullerton (1992) provide a general discussion of the marginal excess burden of taxation in the USA. Findlay and Jones (1982) provide estimates for Australia (see also Chambers (1995) and Corden (1974)). Fox (1985) and Dalrymple (1990) discuss the implications of the deadweight costs of taxation for the measures of benefits and costs of research. References are cited in Alston and Pardey (1996) who summarize much of this discussion and its relevance for financing agricultural R&D.

[21] One can take issue with this position. In an ideal world, with efficient general taxation measures (e.g. income taxes), general taxation revenue would be obtained in the least-cost way, and general taxation would be at least as cheap as a specific commodity tax to fund agricultural research. In such a world, arguments for earmarked commodity taxes to fund specific research programmes would have to follow the general form of arguments for earmarked taxes. In reality, however, least-cost taxation measures seem not to be applied. The issue then is whether, in a general equilibrium setting, at the margin a commodity tax would be a cheaper source of funds for research than the general taxation measures already being applied. In most developed country settings, in which agriculture is relatively effectively subsidized, it could more easily be presumed that a small tax on agricultural commodity production would be more likely to reduce, rather than exacerbate, the social costs of existing distortions. More generally, a view about the particulars of the situation may be required before a judgement can be made about whether commodity check-offs are efficient sources of funds for research. Chambers

(1995) provides a general equilibrium model of the incidence of agricultural policies.

[22] Where research resources are rationed, and the marginal value product of research exceeds the marginal cost of distortions owing to overpricing public goods, it is socially better to overprice the public goods rather than not to produce them at all.

[23] These five countries were Australia, the Netherlands, New Zealand, the UK and the USA: in 1993 these countries accounted for 43% of the industrialized world's public investment in agricultural R&D.

Investment in Natural Resources Management Research: Experience and Issues

12

Gary Alex[1] and Guenter Steinacker[2]

[1]ESDAR, World Bank, 1818 H Street NW, Washington, DC 20433, USA; [2]G.S. GTZ-4233, POB 5180, D-65726 Eschborn, Germany

Introduction

Research on environmental issues and management of natural resources impacted by agricultural production systems is reviewed, with emphasis on the increased importance of natural resources management research. We also identify important issues that affect research investment strategies.

Over the past 20 years environmental and natural resources issues have gained world attention. Environmental movements have focused attention on humankind's stewardship and care for the world in which we live. Desertification, deforestation, species extinction, pesticide residues and watershed degradation have all caught popular attention. Parts of society, especially in industrial countries, have changed their view of agriculture from that of a 'life support system' to an opposite view of agriculture as a major culprit in environmental degradation. This increased environmental awareness has resulted in a major expansion of investment in natural resources and environmental programmes and greater scrutiny of all programmes for their impact on the environment. The expansion in funding for environmental programmes has included increased natural resources management research (NRMR), which has emerged as a major new priority and challenge for agricultural research systems. This expanded research mandate comes at an awkward time, when overall public funding for agricultural research has declined, and when responsibility for research funding is increasingly being shifted to the main users and beneficiaries of research. This can present special problems in the case of NRMR.

Agriculture, Natural Resources and Environment

Agricultural production systems are intimately linked with and dependent on the natural resource base. In many countries, agriculture is the biggest user and biggest abuser of natural resources, including land and soil, water, forests, natural pastures, fish and wildlife. Soils and water are critical inputs to crop agriculture; pastures to livestock production; forests to production of timber, fodder and fuel for rural households; and natural water bodies to water-derived food production. These obvious dependencies have meant that research related to natural resources and their management for productive purposes has always been an important part of the agricultural research agenda.

We use 'environmental' and 'natural resource' somewhat interchangeably, though recognizing that various definitions may present subtle but significant differences.

We will use the term natural resource management (NRM) to refer to agricultural system interactions with natural resources and the environment, including both system use of resources in production and consequences of agricultural production on the natural resources and environment. NRMR will refer to that portion of the agricultural research agenda that deals with agricultural systems' interactions with the environment and the natural resources base. It includes research in biological science, ecology, and social sciences and management systems. Much NRMR may be characterized as 'sustainability research' in that it must maintain a focus on increasing system productivity without permanent degradation of the natural resource base. NRMR may be contrasted with: (i) commodity research including biological research to increase agricultural production and productivity without reference to changes in the natural resource and environmental base; and (ii) policy or systems research relating to social science work to increase agricultural production and productivity without reference to changes in the natural resource and environmental base.

Diverse Stakeholders for NRMR

A wide range of stakeholders is concerned with NRM but, in many countries, those concerned primarily with the productivity of agricultural systems may be in the minority. This is a new development in modern society, when compared to the rural majorities and strong agricultural lobbies of earlier times. The three broad groups of stakeholders for improved natural resources management are: (i) agricultural producers or direct stakeholders; (ii) the 'green' lobby or nature conservationists; and (iii) humankind in general.

Direct stakeholders for NRMR are the agricultural producers and others who rely on natural resources for their livelihood. For these stakeholders, NRMR is critically important and becomes more so with increased costs of natural resources (e.g. irrigation water, which must be pumped further or which is also needed by industry or growing urban areas); decreased productivity of resources (e.g. when rangelands or fisheries are overexploited); potential loss of natural resources for production systems (e.g. loss of land to salinization, erosion, desertification). Fortunately, these losses are quantifiable and economic and provide a basis for stakeholder investment in research to address the NRM issues. Unfortunately, recognition of the true value of resources becomes evident only after resource conditions change and costs increase. Meanwhile, those stakeholders who depend on natural resources for their livelihood frequently lack the organization, awareness of problems and resources to be able to finance essential NRMR.

Improving management of natural resources often means recognizing their true and increasing value. However, recognizing the increased cost of resources, paying these true costs, and supporting research to manage resources in the context of their true value are not high priorities for most direct stakeholders. For example, in the western USA, ranchers jealously guard their subsidized grazing rights on public lands and logging companies lobby hard for access to forest areas. Handling wastes from intensive livestock production systems to avoid nitrate pollution of water supplies is costly, and reducing chemical control options for plant and animal pests can be costly and/or inconvenient to producers. In Asia, cost recovery for irrigation water, which farmers expect to receive free, has been a persistent problem. In Africa, managing livestock and wildlife populations that share the same range may require significant adjustments to livestock management systems. Finally, for many marginal farmers, a long-term concern for the sustainability of natural resource management systems is simply irrelevant, if this means that they will not be able to survive the short term.

A second set of stakeholders is the conservation lobby, which values the aesthetic or social benefits of natural resources. Many of the environmental interests of the conservation lobby are represented through environmental NGOs, which have existed for many years in industrial countries. In recent years, they have grown in numbers and strength and have also become an important factor in the South. These reflect the social valuation of environmental and natural resources and provide strong lobbying groups for natural resource issues. The groups are well organized and, as they often adopt a fairly strict conservationist approach, they have proven to be effective lobbyists and have contributed a large amount to ecological and conservation research. Unfortunately, they have not been as forthcoming in direct funding for agriculturally related NRMR, nor perhaps are they adequately informed as to the value and potential for such research. The conservation lobby draws strength from the fact that natural resources are still to some extent considered free goods, and any scarcity or deterioration is apt to provoke strong responses. People may see it as their right to enjoy nitrate-free, sediment-free water, to enjoy the scenic beauty of natural forests unexploited for fuelwood, or to view spotted owls or elephants in the wild habitat. Thus, agricultural technologies or production systems that impinge on these rights draw quick criticism that may affect public investments in agricultural production and in NRMR.

From another perspective, all of humankind is a stakeholder in NRMR. The environment is truly a life-support system for the globe and affects us all. Global warming, possibly furthered by agricultural practices, may affect all life on earth; the safety of our food supply and drinking water may depend on safe management of agricultural chemicals; biodiversity is thought to be crucial to future applications for human health; and future food supplies depend on sustainable production systems. Stakeholders for these global issues are generally poorly organized. They are, or should be, represented by governments, with occasional public interest group support. The problem is that costs and benefits are uncertain with regard to value, timing and even existence. Governments have responded to these issues, though with a lack of full information and understanding of the problems, and it is difficult to judge whether responses have been adequate or even excessive. The impact of this category of stakeholder on funding NRMR is not yet clear.

These different groups of stakeholders in environment and NRM often have competing interests. Different farmers compete for the same irrigation water. Pastoralists and settled farmers fight over land rights. Slash-and-burn cultivation may affect water resources or fuelwood availability. Interests of direct stakeholders may clash even more with interests of those less directly dependent on natural resources for a livelihood, as when conservationists oppose others who would utilize resources under controlled management. Thus, interests diverge and make it difficult to organize stakeholders to pursue a common interest in NRM or in financing of NRM and NRMR.

Valuing Natural Resources and the Environment

There have long been active programmes of biophysical research on natural resources, as biologists and ecologists have attempted to describe nature and its change. However, debate on the value of natural resources and the environment has recently come to the forefront with the increasing scarcity of natural resources, the growth of the environmental movement and greater environmental impacts from population growth and production system intensification. Values are attributed to natural resources for three reasons which roughly reflect the interests of the three groups of stakeholders for NRM. These values are: (i) the potential productivity and relative scarcity of the resource (economic); (ii) the value of the resource by society (social); and (iii) the potential for irreversible decline in the resource (ecological).

First, natural resources are valued as a basis for agricultural production and for providing people with food, fuel and fibre, and NRMR has long addressed issues of the management of natural resources to maximize production and productivity. The economic importance of natural resources relative to agricultural production systems has increased with their relative scarcity. Natural resources being 'natural' were earlier considered free goods and probably were considered only as factors to be exploited in the production process. Producers would initially have geared their management systems and research investments only towards obtaining maximum production and profits per period of time, irrespective of natural resources used (e.g. maximum crop production regardless of water or land used, maximum livestock production regardless of pasture use, maximum forest product extraction). Later, as the resource base became limiting, focus necessarily shifted towards managing systems to maximize production and profit per unit of resource used (e.g. yield per hectare, production per unit of irrigation water). With further intensification of production and resulting stress on the production base, the focus has begun to shift towards management to maximize production and profit per unit of resource on a permanently sustainable basis. Thus, as natural resources become increasingly scarce, their value and importance in the production system increase.

The second factor affecting the value of natural resources is the perceived value by society. Though most cultures place a high value on maintaining, or at least not needlessly damaging, the environmental base for its staple agricultural production, societies are now placing greater value on a wider range of environmental factors. The social value of natural resources has increased dramatically over the past several decades for two reasons. First, in many societies, but especially in industrial countries, incomes, education and leisure time have increased and relieved much of the population of the simple struggle for survival. This has allowed a greater appreciation of wildlife, clean air and water, and the beauty of the natural landscape. Secondly, as world population increases and production systems are intensified, natural resource depletion and deterioration become more evident, as with pesticide contamination, deforestation and extinction of species. This has led to broad-based environmental movements that place increased social value on the environment. One manifestation has been the placement of blame on agriculture and agricultural research for damage to the environment. NRMR clearly is needed to respond to this challenge.

The possibility of collapse of an ecological system and the absolute loss of natural resources or resource quality is the third factor affecting valuation of natural resources. The potential irreversibility of some natural resource degradation is reflected in species extinction and resulting incalculable loss of biodiversity. Irreversibility is also a factor for forests, pastures or water bodies that become so degraded that they can no longer recover to provide the physical or aesthetic benefits expected from them. Unsustainable agricultural production systems may deteriorate to the point that productivity declines are irreversible and the agricultural systems can no longer provide the basis for a stable economy or an adequate living standard for primary producers.

Integrating economic, social and ecological benefits presents major difficulties for NRMR. Agricultural systems will always have some impact on the natural resource base but, because of system complexities and time lags, understanding and measuring these changes is not easy. To be made comparable and to assess alternatives, natural resource consequences of agricultural production should be valued in economic terms, but it is at this point that the important distinction between natural resources issues and environmental issues arises. Crosson and Anderson (1993a) make a useful differentiation between natural resource and environmental impacts of agricultural systems as follows:

- Natural resource impacts are those consequences of agricultural systems that affect agricultural productivity (land, water, plant and animal genetic resources) and that are usually priced by markets.
- Environmental impacts are those consequences of agricultural systems that affect resources not used in agricultural production and are not usually priced by markets.

Based on this distinction, agricultural system impacts on natural resources would include such things as changes in quantity and quality of irrigation water, soil productivity, changes in quantity and quality of forest areas from which communities obtain fuelwood or other products and changes in quality of rangelands. Examples of environmental impacts would include: changes in quality and quantity of runoff or groundwater not used for agricultural production, changes in land not used for production purposes (i.e. wetlands), changes in species habitats or loss of biodiversity of species of no current productive value and changes in air quality. The distinction between the two categories may blur depending upon the boundaries of the system being observed and the ability to price a resource (e.g. whether or not improved wildlife habitat can be valued by increasing tourist revenues accruing to the agricultural system), or by the location of the agricultural enterprises and relation to other production systems (e.g. whether polluted agricultural runoff is from an upper watershed farm on to adjacent farms or from a lower watershed farm into a city water supply or other non-agricultural use). Valuing changes in natural resources (for which there are markets) is difficult, but problems are even greater with environmental consequences for which no markets exist. These problems of pricing natural resource impacts of production systems are important topics for research and affect how priorities are set for NRMR and how this research is funded.

Changes in NRMR Agenda

The NRM research agenda is quite varied, as reflected in a sampling of research priorities for various categories of natural resource management presented in Table 12.1. Not included in this listing are research topics relating to global environmental issues and the NRMR priorities for basic research to determine the extent to which environmental problems do exist and are influenced by agricultural systems. The fundamental shift in the NRMR agenda in recent years has been from a focus on management of natural resource inputs to the production process to maximize current production and incomes, to a focus on managing these resources for long-term sustainable production. The goal for NRMR is primarily that of developing sustainable agricultural production systems. Research on sustainable agriculture has become important because of the three factors noted above that influence values of natural resources: increasing scarcity of resources, social values on the environment and potential for irreversible decline in ecological systems.

Sustainable agriculture has been defined in many ways. Terms used in the definitions often include: organic, social management, biological, regenerative, ecological, natural, biodynamic, low input, low resources, agroecological and ecoagriculture (Parr *et al.*, 1990). However, the basic element in sustainable agriculture is that it adds a long-term (resource and social) dimension to the assessment of the agricultural production system and seeks an agricultural production system that has a future.

NRM research on sustainable production systems is intertwined with commodity research and to a large extent the two are

Table 12.1. Examples of topics of current high priority for NRMR.

Resource category	Topics of current priority for research
Soil management	Nutrient cycling
	Soil organic matter behaviour and management
	Methods of combating and predicting erosion
	Soil microbiological research
	Managing soil nutrient fertility
	Acid soils management
	Modelling of soils characteristics
Land management	Measuring status and trends in land characteristics (e.g. desertification, vegetation quality, etc.)
	Land tenure research
	Grassland management, including role in carbon sequestration
	Ecoregional classification
	Watershed modelling
	Local government and local organization roles and operations in managing land use
Water resource management	Managing water use for greater efficiency
	Modelling movement of water and nutrients
	Water pricing
	Reuse and recycling of waste water and runoff
	Water management impacts on waterlogging and salinization
Forest resource management	Exploitation of non-timber forest resources
	Agroforestry production systems
	Tenurial rights and management systems for forest use
Genetic resource management	Biodiversity prospecting arrangements
	Ex situ and *in situ* genetic conservation measures
	Conservation of minor breeds of livestock now threatened with extinction
	Park and reserve management and integration with surrounding communities
Ecological system management	Integrated pest management systems
	Relations of agricultural systems to global warming and other global environmental conditions
	Waste management systems for intensive livestock production systems
	Marginal lands management systems
Social sciences	Pricing of environmental and natural resource impacts of agricultural systems
	Social, political and institutional arrangements for natural resources management

Adapted and expanded from TAC (1996) and NRC (1991).

complementary (Crosson and Anderson, 1993b). Plant breeding can reduce the need for agricultural chemicals or water. Improved soil fertility or water management can allow production of improved varieties not adapted to less favourable conditions. There are, however, some important differences and conflicts between commodity research and NRMR approaches. Runge (1992) compares 'environmental research' with traditional 'agricultural research' (Table 12.2).

The agricultural research agenda for NRM is increasingly weighted towards work on sustainable production systems. Factor research for increasing productivity per unit of natural resource (livestock production per hectare of pasture, crop production per hectare, crop production per unit of irrigation water) remains an important goal of much research, but this is increasingly important in the systems context. Sustainability requires a more holistic approach to analysis of agricultural systems and a rationalization of environmental/natural resource impacts with production increases. This holistic approach plus the complexity of NRMR leads to much of it being based on 'systems research'.[2]

Systems research is conducted at various levels. At the highest level, it may look at global production or the global life support system. Examples of this would include work on global warming, global impacts of pollution or resource depletion, or studies relating natural resources and global food production (e.g. Penning de Vries *et al.*, 1995). At the next level down, ecoregional research attempts to define and understand agricultural systems within the boundaries of specific regions, as defined by natural (natural resource and climate), administrative and socioeconomic variables. An important dimension

Table 12.2. Conflicting agenda: comparison of agricultural and environmental research.

Environmental Research + Agricultural Research = Sustainability Research	
Environmental research	Agricultural research
Process-oriented	Product-oriented
Multidisciplinary	Disciplinary
Market failure assumed	Comparatively market-driven
Critical of production as an end	Accepting of production as an end
Resource pessimism in spite of technology	Resource optimism because of technology
Environment valued highly relative to food (North)	Food valued highly relative to environment (South)

Source: Runge (1992).

of systems research is that it provides an effective interface between natural and social science research to improve understanding of physical and socioeconomic constraints on farmers' actions.

Ecoregional analysis allows for interaction of resources within the region (e.g. between forests and communities, nomads and settled agriculture, uplands and lowlands) and provides increased options for development of sustainable production systems. Within the ecoregion, there may be various agroecological zones or agroecological production systems, which utilize different natural resources and/or compete for use of the natural resources. NRMR can seek to increase the efficiency and complementarity of resource use by the different production systems. At the ecoregional level, policy and institutional research addresses the mechanisms for interactions between production systems, which may compete or be complementary in their utilization of the natural resource base. In this context, watershed-level research is an important, simplified approach to ecoregional research (TAC, 1996). The watershed-level research allows:

- oversight of the whole water resource area;
- clear definition of area and identification of externalities;
- definition of uniform land units;
- understanding and testing of current land allocations and uses;
- a framework for spatial analysis of impacts;
- ability to understand and analyse socioeconomic variables; and
- a structure for dealing with biodiversity.

Thus, much NRMR will be systems research, whether global, ecoregional, watershed, community-based, farm-level or field-level.

Another important aspect of the NRMR agenda is that of evaluating and measuring the sustainability of NRM systems. One approach to measuring sustainability is the use of environmental accounting. Sustainability requires, as argued by Crosson and Anderson (1993b), a focus on total factor productivity, considering all costs and all benefits involved in production. This necessitates inclusion of both natural resource and environmental costs and benefits (both difficult to measure appropriately) in any calculation of system efficiency or performance. Alternative approaches recognize that sustainability of production systems requires that the system meet criteria for sustainability in three dimensions: economic efficiency, social equity and ecological sustainability. The interrelations of these three aspects of system performance provide the complexity in much of NRMR (Mueller, 1996). Mueller advocates use of an indicator system (see below) to measure sustainability of agricultural systems.

Policy research, with either a commodity focus or an NRMR focus, can have immense impact on natural resource use, chiefly by affecting prices and valuation of resources. Consequently, policy research, especially as it relates to access to and pricing of resources, is important for NRM research.

NRMR Institutions and Financing

NRMR funding policies of stakeholders

The considerable increase in attention to natural resource and environmental issues has resulted in substantial increases in funding for such research. This has been one of the few growth areas for agricultural research in recent years. The increased financing has come from both the pressures from interest groups and from scientific concerns over issues of national or global importance.

Governments
Governments collectively have accorded the environment a high priority in international fora. The 1992 UN Conference on Environment and Development established an ambitious environmental agenda for the international community. This has been reinforced by other conferences or agreements on tropical forests, biodiversity, desertification and others. These international agreements have effectively focused attention on environmental issues, though this may not always have carried over to action and programme financing by individual countries. Public investments may tend to focus on basic research to understand global problems, e.g. global warming, desertification and conservation of biodiversity, and research on problems of high profile or local interest, such as a polluted river, deformed frogs or a specific endangered species. There is not the same enthusiasm for sustainable agricultural systems work, though this NRMR funding has also increased.

In centrally planned economies, lack of response to popular pressures has led to neglect of environmental controls and research, and left many environmental problems unattended. In these and other areas, an important focus for research will be on natural resource tenure and land use systems.

NGOs
The NGO and environmental lobby has performed a valuable service in raising consciousness and understanding of natural resource and environmental issues. Many individuals in these organizations have been scientific and professional leaders in the conduct and promotion of NRMR. The nature of environmental NGO support and fund-raising causes some difficulty in supporting a balanced NRMR programme.

NGOs depend largely on private donations, which are most easily mobilized for activities with an immediate impact or with a high profile. Thus, some NGOs tend to support immediate and popular implementation projects, rather than the necessary long-term research that may provide direction to such projects.

Private sector
Private industry financing for NRMR has increased, but is quite variable in coverage and motivation. Some companies have invested in NRMR for public relations reasons or to stay ahead of environmental regulations and the environmental lobby. Environmental legislation and regulations can provide major incentives for private sector investments in research to reduce negative environmental or natural resource impacts of agricultural systems (e.g. use of agrochemicals or improved handling of livestock and crop wastes).

A natural resource of major interest to the private sector is that of genetic resources, for which substantial private funding has been available for conservation and prospecting. Genetic resources range from those of immediate value, as for some major crops and related plant species, to 'potential' value for the genetic resources locked in other species in the tropical forests and elsewhere. Private financing for research on genetic resources is forthcoming from seed companies and biotechnology companies, but only for targeted cases. Public funding for research on genetic conservation strategies and technologies has become a relatively high priority, though much remains to be done in this area. Valuing of genetic resources and developing parameters for their use and property rights remain high priorities and present difficult problems. The institutional roles and implications for funding of NRMR by different stakeholders are summarized in Table 12.3.

Impact of Changing Funding Policies on Research Institutions

Research institutions have responded to the new emphasis on NRMR in various ways. The CGIAR has established 'protecting the environment' and 'saving biodiversity' as two of its five objectives, and in 1997 devoted approximately 34% of its non-policy research funding to these objectives. The CGIAR system now includes four NRM centres: IIMI, ICLARM, ICRAF and CIFOR. The CGIAR budget trend reflects the changing priority for NRMR. NRM research has shifted from 29% (US$50.8 million) of the total CGIAR budget in 1993 to 45% (US$122.7 million) proposed by TAC for 2000 (CGIAR, 1997a). It has also introduced agro-ecoregional programmes and integrated systems research

Table 12.3. Roles of stakeholders in funding NRMR (sustainability research).

Stakeholder	Role of stakeholder	Likely research investments	Other interventions
Society at large (through governments)	Ensuring long-term public goods perspective Balancing sustainable development (growth, equity, ecology) Ensuring conservation of critical environmental resources	Basic research to understand global problems and agriculture–natural resource interactions Research on tenurial and other policy issues Complex systems research	Establishment of legal and policy frameworks for agricultural systems, including incentives for appropriate NRM and investment in NRMR Convening national and international meetings on key issues Developing NRMR capacities
Environmental lobby (through NGOs)	Develop public awareness Lobby for additional NRMR funding	Research on conservation of targeted environmental and natural resources Research on monitoring of environmental conditions Research on systems for community-management of resources	Education and information activities related to NRM Defining social values for environmental issues Lobbying for NRMR Monitoring of research and production activity impacts on the environment
Private sector (farmers)	Increase production on a sustainable basis Ensure maintenance of resource base for production Maximize profits on investments	Research on conservation of critical elements for the production process (only when threatened) Research on efficiency of use of high-priced inputs Research on alternative production systems	Lobbying on government policy for NRM Conservation of natural resource base for production

continued

Table 12.3 continued

Stakeholder	Role of stakeholder	Likely research investments	Other interventions
Private sector (other than farmers)	Maximize profits on investments Increase production on a sustainable basis Ensure maintenance of resource base for production	Research on genetic resources Research on equipment or patentable techniques for more efficient use of high-cost natural resources Research on minimizing environmental effects of production or processing industries, when controlled by government regulation	Lobbying on government policy for NRM Conservation activities for public relations

programmes, such as the African Highlands Initiative and many individual centre programmes for NRMR.

The shift in funding priority to NRMR has exacerbated budgetary stresses, as the environmental emphasis in research programmes may not have brought greater resources to research programmes. Total funding has grown slowly, if at all, though new NRMR programmes have been added. Of USAID funding for the CGIAR in 1996, 25.2% went to CGIAR NRMR programme categories, though only slightly over 1% of the USAID contribution derives from funding for environmental programmes and the balance is from an account for economic growth programmes. With their limited funding, donors have emphasized NRMR even in more traditional commodity programmes in order to maintain programme support and reflect current political priorities. Research institutions have also followed this lead and have attempted to justify funding and portray many programmes as environmentally related.

Through the 1980s the CGIAR struggled with the conflicts between increasing productivity and sustainability of agricultural systems, but has now gone far in integrating sustainability and NRMR into its research agenda (Walsh, 1991). Some shift in funding was clearly needed, though the balance between commodity and NRMR is a matter for continuing debate. The priority given to NRMR is also reflected in the USA, where the Land Grant Colleges and Universities place 'environment and natural resources' first in their 1994 listing of six main areas of agricultural research (NRC, 1995). Key elements of this research agenda are: conservation and enhancement of natural resources; management of ecosystems to conserve biodiversity; recovery and use of waste resources through agriculture and forestry systems; and development of resource management decision systems. In 1992, 18.1% of research expenditures by the US Land Grant Universities was for the Environment and Natural Resources programme area and 18.2% of scientists worked in this area. In 1992, 24.6% of US public research expenditures were for natural resource-related research, including forest resources (13.1%) and natural resources (11.5%) (Fuglie *et al.*, 1997). From 1972 to 1992, NRMR expenditures grew steadily, increasing from 12 to 15% of total US Department of Agriculture (USDA) research funding.

Thirty out of 75 US agricultural colleges and universities explicitly recognize the importance of natural resources in their mandates by their names, which have changed from 'college of agriculture' to 'college of agriculture and environmental sciences' or some other variation on this theme. Environmental science programmes have expanded greatly with many different institutions involved in research, some related to agriculture and natural resources in addition to other environmental issues. This trend has been general in industrial countries, and in Germany some agricultural research institutes have completely shifted their mandate to cover environmental issues.

However, in the USA, the research system has been criticized for being slow to address environmental concerns (Fuglie *et al.*, 1997). In response to this criticism, the 1990 Farm Bill directed the USDA to ensure that competitive grant funding of research is consistent with development of sustainable agricultural systems. In practice, it has been difficult to comply with this requirement, because of the problem of balancing attention to the economic and environmental aspects of sustainability. The criticism of research institutions for being slow to respond to environmental concerns has resulted in a shift from fixed formula funding for research programmes to greater use of competitive grants funds, which can be more easily used to direct shifts in research focus.

From 1978 to 1992, NRMR conducted by USDA agencies declined (to US$267 million), while NRMR by state agricultural experiment stations (SAESs) and other research institutions increased (to US$465 million). Over this period, USDA funds for water, land management, pollution and forest products declined; soil research grew slightly; and 'other' research, including weather, remote sensing and interdisciplinary research grew substantially. At the SAESs and other institutions, 'other' research, especially interdisciplinary research and fish and wildlife research, are the areas receiving the most funding. State tax revenues provide an increasingly important source of funding.

National agricultural research systems (NARS) in developing countries have not shifted programmes towards NRM with as much enthusiasm as have industrial countries. Though NARS are influenced by donor funding priorities and by increased awareness of natural resources issues, there is, to some extent, still a view that environmental issues are 'rich country' issues. For Africa, due to the high level of donor support for research programmes, the growth of NRMR is thought to be relatively high, though this donor dependency may make the sustainability of programmes questionable. Overall, the slow change in many NARS research agenda may also be due to: lack of capacity in the NARS to deal with natural resources issues; institutional rigidities in many of the NARS; and conservatism of many programmes as to allocation of funds by scientific discipline.

Lessons Learned

Recent experience with NRMR suggests the following lessons for financing of research.

Importance of environmental accounting

The key to addressing environmental and natural resources issues lies in incorporating the costs and benefits from natural resource and environmental impacts into consideration of agricultural sector performance. In national accounts, *green accounting,* which incorporates costs and benefits of changes in natural resources and the environment, can draw attention to the importance of changes in these factors and the sustainability and real growth rate of an economy. There are formidable challenges in quantifying environmental changes and assigning economic values to natural resource and environment impacts, and much needs to be done before environmental accounting is practical.

Individual production systems also need to be evaluated on the basis of environmental and natural resource costs and benefits. The key to ensuring that production systems are sustainable and that there are appropriate investments is to internalize environmental externalities; that is to ensure that such production systems pay all environmental costs and are credited with all environmental benefits. Natural resource issues (which directly affect the production system) may be fairly straightforward to address in NRMR. For example, pricing of irrigation water provides an immediate and direct incentive for producers to make most efficient use of water. This provides further incentives for research on irrigation technologies, irrigation management and production systems to make most efficient use of water resources. Similar cases could be cited for pricing forest resources, pasture lands or fishery resources.

Crosson and Anderson (1993b) argue for use of environmental accounting in evaluating alternative research investments and in prioritizing agricultural research programmes for NARS. In theory this may be correct, but data and the methodologies for valuing costs and benefits may be inadequate for this to be practical in many countries at this time. Development of simplified systems for such prioritization of research projects may be an appropriate step in this direction. Regardless of whether environmental accounting is used for evaluating production systems and proposed research investments, it is clear that correct pricing of the environmental consequences of agricultural systems and the formulation of policies for appropriate allocation of costs and benefits of those environmental consequences will be central to addressing NRM problems. This leads to the further observation that policy research capacity and economic expertise will have to be greatly strengthened in many research systems to address these issues.

Importance of research approaches

NRMR requires considerable integrated systems research approached at national, ecoregional, community and farm levels. TAC (1996) lists the following ingredients for success for NRMR policy and management research:

- Participation of local organizations.
- Use of specially trained catalysts for local work.
- Use of interdisciplinary and interinstitutional approaches.
- Adoption of a learning process mode.
- Attention to appropriate technologies.
- Use of horizontal diffusion of innovation.
- Building on indigenous institutions.
- Exploitation of outstanding leadership.
- Bureaucratic reorientation of implementation institutions.

This illustrates the importance of social science input to NRMR. In practice, this needs to be complemented by biophysical research,

though the integration of these disciplines presents significant institutional and methodological difficulty and may require reallocation of
traditional funding. There is therefore need to integrate public funding, possibly from different sources, for research by different disciplines
on common agricultural problems.

The answer to the problem of organizing the interdisciplinary
research needed for NRMR is generally not in establishing new institutions (Scott, 1992; Goldsworthy *et al.*, 1995), but in facilitating interactions and research around a common theme. Strategies to accomplish
this might include: establishing a lead agency, creation of a special
project unit, parallel financing of different institutions or joint financing (ISNAR, 1995). Examples of the various efforts at organizing such
interdisciplinary work on NRMR issues are outlined below.

1. CGIAR ecoregional initiatives. The CGIAR initiated approximately
ten ecoregional or collaborative NRMR-related initiatives (Goldsworthy
et al., 1995). These included: the Consortium for Sustainable Development in the Andean Region; Alternatives to Slash-and-Burn; Lowland
Rainfed Rice Consortium; Rice–Wheat System Consortium; On-Farm
Water Husbandry in West Africa and North Asia; Coastal Environments;
Humid and Sub-humid Tropics of Asia; Desert Margins Initiative; and
others. The definition of these programmes was not always entirely
clear, though this has improved over time. These initiatives appear to
have experienced significant difficulties in start up. By April 1996, of
the seven ecoregional programmes proposed in 1993, four were still
partly or entirely in the design phase, two were operational and one
had been incorporated into another programme. Since additional funding was often not forthcoming for these programmes, in some cases
implementation lagged or established programmes were redefined to
fit as ecoregional research. Apportioning funding can be a problem in
such interinstitutional initiatives and this is complicated by problems
of allocation of programme funding between direct research costs and
the overhead costs and by high transaction costs for individual
institutions.

2. Another example of a multiple institution approach to addressing
an ecoregional problem is the African Soil Fertility Initiative, an attempt to understand and reverse the decline in productivity of African
soils. This initiative, which is still in the initial planning phase, is
aimed at influencing investments and programmes and impacting on
sustainable fertility and productivity of African soils. It has been
successful in bringing together a wide range of private and public
sector institutions involved with the sector. It encourages country-
level action plans to address the decline in soil fertility and focuses on
policy, technology and institutional changes necessary to improve soil
fertility conditions.

3. The Land Quality Indicators (LQIs) programme is a global coalition of international and national institutions led by the World Bank, FAO, UNDP and UNEP to develop a system of land quality indicators to monitor the status and change in quality of land resources with land defined as 'all elements of the earth's terrestrial surface that affect potential land use and environmental management.' The LQIs are based on the pressure–state–response framework, with indicators reflecting pressures on land resources, changes in the resources and social responses to these changes. The programme has a donor support group for funding, a core advisory committee for scientific guidance and a secretariat to coordinate activities.

4. Beginning in 1992, USAID funded a Sustainable Agriculture and Natural Resources Management (SANREM) Collaborative Research Support Programme (CRSP) to implement a comprehensive, farmer-participatory, interdisciplinary research programme to elucidate and establish principles of sustainable agriculture and natural resources management. A 1996 management review of the SANREM CRSP concluded that the programme should be extended, but found that the need to involve users in all aspects of programmes in order to develop methods and practices for sustaining natural resources had been time-consuming. The review noted the 'novelty and complexity' of the programme and, though the review was generally positive and the programme is continuing, there have been some funding constraints (Hamilton *et al.*, 1996).

5. One of the most successful examples of NRMR has been that of IPM for rice in Asia. In Indonesia, the often-quoted success story, the programme's success seems to have been based on initial research on the brown planthopper, a clear problem, government policy change and sustained multi-donor support (Kenmore, 1996).

Success with these programmes seems to have been, in cases such as Indonesian IPM and the Indo-Gangetic Plain Rice–Wheat Consortium, where the systems research is quite focused on a production problem and commodity within the production system. In addition, the general experience with these programmes seems to be that management and coordination costs for integrated, multi-institutional NRMR programmes are substantial and that best practice for effectively organizing and funding such integrated programmes has not been well established. In addition to inherent difficulties with organizing multidisciplinary research, the NARS of developing countries are also hampered by a lack of capacity to undertake NRMR, in both socioeconomics and the necessary biophysical sciences.

Competition with commodity research programmes

Some of the difficulties with the emergence of NRMR and its integration into overall research programmes are likely to be a result of competition with the established commodity research programmes. These two types of research should be complementary, but in a declining budget environment, commodity research programmes cannot be expected to welcome new competition for funding. This problem may be exacerbated by the following:

- Crop genetic improvements which provide benefits that are immediately visible, whereas NRM sustainability research may provide impacts only in future production.
- NRMR frequently requires support from social sciences and other disciplines not formerly prominent in agricultural research.
- The complexity of much NRMR systems work, which makes it difficult to explain and slow to produce results.
- Changes in resource management practices are frequently more difficult for farmers than are changes in a single component technology.
- Difficulties in start-up of many of the multi-institutional and multi-disciplinary programmes.

Measuring impacts and allocating resources in NRMR

Measuring impact of research programmes and setting of research priorities (essentially an *ex ante* impact assessment exercise) present difficult challenges in the best of situations (Anderson, 1997). These challenges are further complicated by the multiple objectives inherent in most NRMR. Since valuing all impacts in economic terms is difficult, and frequently not possible, various indicators may be needed to evaluate research impacts. Considerable effort has gone into identifying and developing appropriate indicators to measure environmental impacts of agricultural projects (OECD, 1997). At the level of individual research activities, balancing priorities is key to developing sustainable agricultural systems. The balance must be between equity, economic efficiency and ecological sustainability of the system. Lack of economic efficiency ensures that the system will fail and be replaced by a more efficient system; lack of social equity and the institutional structure for the system may be threatened; and lack of ecological sustainability will eventually bring down the system.

Mueller (1996) notes that indicators are needed for analysing sustainability of systems and can make sustainability a more operational concept. Such indicators should reflect balance between equity, environment and efficiency of a system. Thus, systems such as that proposed by Mueller use large numbers of indicators to cover all aspects of sustainability of a production system. This introduces problems of assigning weights to and balancing the various indicators. Considerable research is needed to develop methodologies and frameworks for use of indicators in measuring system sustainability.

Allocation of funding for NRMR will probably remain quite a political art in balancing the three dimensions of system sustainability, which interact with and – to some degree – interact through different stakeholders. Farmers and private sector stakeholders will emphasize research on economic efficiency, both in commodity research and NRMR. Though this may be tempered by concerns for ecological and social stability of systems, economic efficiency will be their most immediate concern. Donors and governments, on the other hand, may be most concerned with equity (through economic efficiency, where possible) to maintain the institutional stability and sustainability of systems. Ecological sustainability (NRMR), though critical to system sustainability, has a less clearly defined or stable constituent base. Considerable future research will be needed to develop more precise indicators, refine procedures for data collection and quantify relationships between indicators and objectives (OECD, 1997).

NRMR linkages to policy formulation

To a greater extent than with commodity research, NRMR tends to be of a public goods nature and is dependent on government policies and regulations to utilize research results and to appropriate benefits from the research. This requires that NRMR maintain close linkages with government policy-makers. This was strongly endorsed at a 1996 workshop (Preuss, 1996). In that workshop, Springer-Heinze (1996) emphasized the two-way relationship between NRMR and public policy. Effective NRMR depends on appropriate public policies to provide incentives for adoption of new technologies developed, especially for innovations to reduce negative externalities associated with agricultural systems. At the same time, NRMR is needed to inform public policy formulation as to the potential natural resource or environmental costs and benefits associated with various policy options. NRMR needs to be proactive in identifying and pushing policies that will have positive impacts on the environment and that will foster adoption of improved NRM technologies and practices. NRMR is therefore more frequently and more closely associated with and more dependent on policy and political structures than is commodity research. This appears true at all levels: international, national, community and family.

Implications for NARS

The above discussion of NRMR suggests the following for consideration by NARS and donors in sustainable financing of NRMR in agriculture:

Valuing natural resources and environmental impacts

Pricing is important. Government policies that correctly allocate costs and benefits of environmental impacts will provide direct incentives for conservation of resources and also for financing of research on related NRMR issues. Without such incentives, there will probably be little interest in private sector investments in NRMR. Thus, research on pricing and allocating natural resource costs and benefits should be an important part of the NARS research agenda.

Promotion of NRMR stakeholder groups

Stakeholders for NRMR are not well or evenly represented. NGOs, especially, but also governments and producer groups, need more information on the role of NRMR and its importance to their interests. This may come through training and also through involvement as stakeholders in the research planning process. This may help to increase the legitimacy of the groups and their credibility in lobbying for research. It may also be a means of tapping an additional source of financing for research.

Funding for multidisciplinary systems research

Much of NRMR must be multi- or interdisciplinary and organizing such can be complex. It is usually not advisable to establish new institutions for such work, but to organize programmes that draw research input together from various specialized institutions. Research issues are complex and may have better potential for success if broken down into fairly well focused objectives. Financing such interinstitutional programmes can be as complex as the research itself. It may be best to ensure that programme-funding decisions are taken outside of the specialized institutions to avoid NRMR funding being used for commodity research masquerading as an NRMR programme. Research networks have proved to be effective for coordinating research across institutions and may be a useful model for NRMR.

Capacity building for NRMR

Many research systems are weak in social sciences, economics, ecology and some of the basic biological sciences needed for effective NRMR. This will require some additional training and reorganization to use staff effectively in these disciplines.

Natural resource and environmental databases

Much NRMR requires an extensive foundation of data for decision making and for understanding environmental impacts. This is needed as a base for developing monitoring indicators and for assessing natural resource impacts. Systems research must integrate these data sets between different scales (farm, community, watershed, national, ecoregional). Geographical information system technologies, improvements in remote sensing technologies and current work on indicators will aid this work, but countries will have to invest in data sets and data management systems.

Notes

[1] Natural resources are defined as 'productive materials and capacities supplied by nature'. The environment is the 'complex of physical, chemical, and biotic factors (as climate, soil, and living things) that act upon an organism or an ecological community and ultimately determine its form and survival'.

[2] To some extent, the evolution of this sustainable systems focus is mirrored in the shift in research focus over the past decades from cropping systems research to farming systems research to community-based resource management, and to ecosystems analysis and monitoring economic impacts of natural resource changes.

New Biotechnologies: an International Perspective

13

J.J. Doyle[1] and G.J. Persley[2]

[1]45 St Germaine, Bearsden, Glasgow, UK; [2]Biotechnology Alliance Australia, University of Queensland, St Lucia, Brisbane, Australia

Introduction

The impact of modern biotechnology on the world's economy is becoming increasingly evident, as the past two decades of investments are now resulting in a wide range of new products, processes and services. Total sales of these new biotechnology-based products in the USA, presently the major market, were approximately US$8.8 billion in 1995. It is estimated that US sales alone will increase by 12% annually over the next decade, leading to total sales of approximately US$31.4 billion by 2006 (Ernst and Young, 1995).

Biotechnology has been defined by the United States Office of Technology Assessment (OTA, 1991) as 'any technique that uses living organisms to make or modify products, to improve plants or animals, or to develop microorganisms for specific uses'. It comprises a continuum of technologies, ranging from *traditional biotechnology* based on long-established and widely used technologies, such as those used in fermentation, through to *modern biotechnology* based on more recent techniques of recombinant DNA technology to enable the genetic manipulation of living organisms, modern immunology as a basis for new diagnostics and vaccines, and new cell and tissue culture techniques for the production of biological products. Biotechnology is thus not an industry in itself, but a set of enabling technologies which are being applied to research and product development in several existing industries, notably in the pharmaceutical industry, agriculture and food processing and in the conservation of the environment.

International Perspective

Evolution of technology and investment

Modern commercial biotechnology originated in the USA in the mid-1970s when the business and scientific community recognized the commercial opportunities arising from the new discoveries in genetics, immunology and biochemistry. Biotechnology has been described 'as the first business with enough glamour to persuade eminent scientists that the entrepreneurial spirit and academic respectability are not mutually exclusive' (Wyke, 1988).

The major reason for the greatly increased role of the private sector in modern biotechnology is that for many of the new technologies the process and/or the product are protectable by patents and other forms of intellectual property rights. Thus a company is able to appropriate many of the benefits of its R&D investments in biotechnology, in contrast with previous public good research in biology, from which an individual or organization could not benefit directly from inventions by the commercialization of the intellectual property involved. This new situation means that powerful new discoveries in biology and genetics are able to be developed into valuable commercial products.

In the USA, biotechnology developed as a business initially through the creation of new biotechnology firms, some 400 of which were already established in the USA by the late 1970s. The characteristics of these start-up companies are that they are small, technology-based enterprises, often established in close association with a university. They are usually dependent on venture capital and equity investments. They have a strong R&D component and a long lag time to product development. Many of the new biotechnology firms are loss makers, at least initially, and often need to go to the capital market more than once for venture capital and equity investments.

The major companies in the pharmaceutical industry and agribusiness have also progressively utilized biotechnology in their businesses, in order to develop new products, to reduce the costs of production, to improve productivity or to conserve the environment. The areas where the major companies are investing in biotechnology are in human health care for new diagnostics, drugs and vaccines; and in agriculture for products such as new biopesticides and novel crop varieties with improved characteristics, such as improved pest resistance, extended shelf life or better processing quality. Approximately four times as much has been invested in R&D for biotechnology applications for human health care as for agriculture. Total biotechnology-related R&D expenditure in the USA in 1995 was US$7.0 billion (Ernst and Young, 1995).

The major companies became involved in biotechnology through both in-house R&D and contract R&D with academic institutions and new biotechnology firms, from which the major companies owned or shared any resulting intellectual property. There has also been a significant public sector investment in modern genetics and related biological sciences, through universities and national research institutes, notably in the USA, Japan, Australia and several European countries through the 1970s and 1980s, building on the discoveries in recombinant DNA technology and immunology of the early 1970s.

The biotechnology business evolved differently in the USA and Europe, in that approximately two-thirds of all new biotechnology firms were established initially in the USA. Relatively few were established in Europe, where biotechnology has developed primarily through the major European pharmaceutical, chemical and food companies, especially those operating globally from France, Germany, Switzerland and the UK. Similarly, in Japan, the large Japanese companies are the major investors. Australia has a small number of start-up, biotechnology-based companies as well as activity by larger companies, both Australian-owned companies and multinational companies and their Australian subsidiaries.

Changes in the International Environment

There have been several significant changes in the international environment that have led to consolidation and restructuring within the biotechnology sector. As a result of the 1987 stock market crash in the USA and elsewhere and the consequent loss of investment capital, many small, public biotechnology firms either merged, were acquired by larger companies or went out of business. The constraints on the early start-up companies were: that they lacked sufficient development capital; it took longer and cost more for product development than originally expected; regulatory processes for product approval were slow and costly; and the new firms lacked marketing and distribution channels for new products. Because of these constraints, most new biotechnology firms that did develop promising technologies and/or products have entered into strategic alliances, joint ventures or been acquired by major companies, to ensure the successful commercialization and distribution of their products. Having such partners also became important in moving new products through the regulatory channels, a process with which the pharmaceutical and chemical companies were well experienced.

The regulatory process for pharmaceuticals developed using recombinant DNA technology is essentially the same as for conventional drug development, based on chemical technology. In agriculture, the development of suitable regulatory arrangements has been more controversial, owing to perceived threats to the environment and to biodiversity by the release of genetically engineered organisms. The US system is based largely on assessing the familiarity of the product and its characteristics. The European system places more emphasis on the process by which it is produced. A safe and efficient national regulatory system with facilities that meet internationally agreed regulatory standards is a comparative advantage for countries investing in biotechnology (Persley et al., 1992; Doyle and Persley, 1996).

Market Size and Trends

In the 1990s there has been a new wave of investment in biotechnology as the first suite of novel products has come to market successfully. Total industry sales in the USA were approximately US$9.1 billion in 1995, an increase of 21% over the previous year. The estimated sales for 1996 were US$9.8 billion. It is estimated that these sales were for human therapeutics (75%), human diagnostics (17%), agriculture (3%), specialities (3%) and non-medical diagnostics (2%). An annual average growth rate of 12% is anticipated over the next decade, leading to a market of approximately US$31.4 billion in the USA alone by 2006.

Pharmaceutical industry

The early biotechnology-based products have been mainly human therapeutics. Product sales in 1996 were expected to reach US$7.8 billion. There are presently 34 products on the market in the USA. These are mainly products for gene therapy and new vaccines. There are approximately 284 potential new pharmaceuticals in clinical trials in the USA, 40% of which are being tested for cancer treatments and 10% being tested against AIDS/HIV (PHARMA, 1996). Biotechnology is now an integral part of new drug development. It is estimated that by the year 2000, 85% of all new pharmaceuticals will be produced as the result of using biotechnology in both the R&D and production processes. New diagnostics is an area where new biotechnology companies developing products for niche markets have been especially successful. Diagnostic monoclonal antibodies are being used in drug testing, cancer detection, pregnancy and fertility testing, and the diagnosis of infectious diseases such as AIDS and herpes. Sales of human diagnostics are expected to be US$1.75 billion in 1996 and grow at least 9% annually.

Agriculture and food processing

Biotechnology now has increasing commercial applications in agriculture and food processing. There are at least 15 new products on the market, with the first releases of new varieties of crop plants with novel traits (e.g. cotton with insect resistance, potato with disease and insect resistance, and tomato with extended storage life; canola with herbicide resistance). In 1996, it was estimated that there were approximately 0.6 Mha of novel crop varieties grown in the USA. Other products on the market include new diagnostics for plant and animal diseases, biopesticides and vaccines to prevent animal diseases.

Annual sales of the novel agricultural biotechnology products were estimated to be US$277 million in 1996, with a rapid annual growth rate of 20% predicted over the next decade as the many new products in the pipeline come to market. In the agriculture and food sector, it is recognized that the products of biotechnology will be delivered mainly through seed. The time to commercial release of some products has been delayed by the lengthy and controversial regulatory processes in the USA and the European Union, due to public perceptions of possible risk from the release of genetically engineered organisms into the environment. The regulatory system in the USA and Canada is being streamlined but the EU system continues to be time-consuming and controversial.

Environmental Biotechnology

Biotechnology techniques, based on the use of microorganisms, can be used to clean up hazardous waste sites, to degrade industrial waste, to clean up oil spills and to rehabilitate mining sites. While long-standing problems in industrial countries, these environmental problems are also emerging in the newly industrialized countries and new, biologically based solutions are being sought. Environmental biotechnology is a growing segment of the market, with sales forecast to grow from US$219 million in 1997 to up to US$584 million by the year 2000. Biotechnology also offers the possibility of more environmentally sustainable agricultural production, for example by the use of novel biological agents rather than pesticides to control plant pests and weeds, an advantage both to the environment and to the consumer. There are several new biopesticides emerging, with potential for use in integrated pest management systems in a range of crops where pesticide use is presently excessive and damaging to human health and the environment (Waage, 1996).

Production and marketing

The novel products arising from biotechnology have been taken to market mainly by the major pharmaceutical companies and by the agrochemical and seed companies. Many of the technologies were developed either by public sector laboratories or by new biotechnology firms. These technologies have either been licensed to the major firms or the latter have acquired new firms with promising technologies. The major companies have three advantages: (i) a global distribution and marketing network; (ii) sufficient cash flow to acquire new firms with their intellectual property; and (iii) the funds to continue to reinvest a proportion of their profits in R&D to develop the next generation of novel products. This trend towards mergers and acquisitions is particularly evident in agriculture, where much of the enabling technology, intellectual property, patents, knowledge and investment in the commercial use of agricultural biotechnology now lie with a small number of multinational companies. Access to, and freedom to operate with, the core technologies held by these firms in order to evaluate their applicability to other commodities and problems worldwide is a critical issue in agricultural biotechnology.

Investment Trends

Since 1995 there has been a new wave of private investments in biotechnology as investors try to capture the benefits of the new technologies. Ernst and Young (1997), in their annual review of biotechnology, note that there were approximately 1300 biotechnology companies in the USA, about 650 of which were small biotechnology companies and 260 were large public companies, mainly in the pharmaceutical, chemical and food industries. There were a further 485 biotechnology firms in Europe, mainly large companies in the UK, France, Germany, the Netherlands and Belgium.

The major companies are continuing to invest heavily in contract R&D with universities and small biotechnology firms and owning or sharing the intellectual property if any potential commercial products or processes emerge. The new biotechnology firms continue to seek strategic alliances with other companies for product development and distribution. The number of such alliances in the USA increased from 152 to 246 between 1994 and 1995, a jump of 62%. Approximately 30% of these alliances were with overseas partners (Ernst and Young, 1995). Venture capital remains a major source of funds especially for small and medium-sized biotechnology companies. Venture capital-financed biotechnology companies in the USA raised an average of US$4.7 billion per year in private equity finance from 1992 to 1994.

The trend in the USA since 1994 is for venture-capitalists to invest in mature companies when they have developed promising products to a stage close to commercial marketing rather than start-up companies (Ernst and Young, 1995).

Another trend is the restructuring and reduction in public sector expenditure in science in several OECD countries, with greater emphasis on public sector institutions earning a proportion of their income from contract research and the commercialization of their inventions. Several OECD countries have various tax and financial incentives to encourage private sector participation in biotechnology. One of particular interest is the UK initiative on *Biotechnology Means Business*, to which the UK government has committed £10 million (US$16 million). The programme, managed by the Department of Trade and Industry, aims to inform companies of the opportunities and benefits of using biotechnology in their business, with assistance provided through a telephone help line, follow-up technical advice, seminars, publications and competitive grants, including a new grants scheme to encourage public sector agencies to collaborate in assembling portfolios of intellectual property suitable for commercialization (DTI, 1996).

Several other member countries of the European Union are also promoting biotechnology, through promoting the growth of existing and new biotechnology companies, direct government support, the encouragement of venture capital investment and the adoption of less restrictive regulatory systems. The UK presently leads Europe in terms of the number of biotechnology-based companies and the promotion of an active financing environment. France and Germany have also begun to examine ways in which they can stimulate the development of local biotechnology-based companies as they recognize their potential impact on employment and sustainable industrial development.

Another changing factor is that several developing and newly industrialized countries are now investing substantially in biotechnology for its applications in human health care, sustainable agriculture and food production. For example, there is increasing interest in investments in biotechnology in Asia, especially in China, India, Korea, Malaysia and Singapore. The Malaysian Government has established a National Directorate of Biotechnology, with substantial funds to invest in the development and commercialization of biotechnology. Similarly, Singapore has substantial public and private funds to invest in biotechnology as well as various tax and other financial incentives to encourage biotechnology companies to establish operations in Singapore. The development banks, such as the World Bank, and a number of bilateral agencies and private foundations are assisting many developing countries through grants, loans and technical assistance to develop capacity in biotechnology. The countries include Argentina,

Brazil, China, India, Indonesia, Kenya and Zimbabwe, among others. Biotechnology is seen as being essential for increasing incomes of people in poorer countries and meeting world food needs into the next century (CGIAR, 1997b).

The World Bank has lent at least US$100 million for biotechnology-related activities. Bilateral development agencies, such as those of the USA, the UK and the Netherlands, and private foundations such as the Rockefeller Foundation, have invested approximately US$200 million in biotechnology R&D over the past decade (Brenner, 1996). The international agricultural research centres, sponsored by the Consultative Group on International Agricultural Research (CGIAR), a group of some 50 governments and private foundations, presently spend approximately US$22.4 million per year on biotechnology R&D for crops and livestock important throughout the developing world (CGIAR, 1996b). The CGIAR is presently examining ways to increase its involvement in biotechnology (CGIAR, 1998a,b).

Emerging Technologies

The emerging technologies that will have most impact over the next decade are: molecular genetics, particularly genome mapping; cellular and molecular immunology, including the role of cytokines in human health and the development of novel vaccine technologies; and new technologies for drug design and synthesis based on knowledge of the structure and function of the drug and its target. The International Human Genome Project aims to map the entire human genome, at a cost of some US$3.2 billion over the next decade. This project is likely to lead to a new generation of human therapeutic products as the genetic basis of various diseases is elucidated. The genomes of economically important livestock and plant species are also being mapped through international collaborative efforts. Such information will improve the rapidity and accuracy of the breeding of new varieties and breeds of plants and animals with improved performance traits. It will also support the production of transgenic plants and livestock with improved performance characteristics not attainable by current breeding technologies.

Epilogue

Summary and Next Steps 14

Introduction

The lessons learned from the country and regional studies are summarized in this chapter, and the issues and options across key themes are synthesized. Also covered is: experience to date, what has worked and why, and ways to identify some of the potential problem areas for those countries/institutions embarking on major structural and financing reforms.

Forces for Change

Support for agricultural and natural resources research is changing worldwide, with a trend towards decreasing public sector investments in R&D in most countries. This is partially offset by increasing private sector investments in agribusiness, including R&D investments aimed at the generation of new products and processes resulting from the applications of biotechnology. There is also growing interest in participatory processes involving farmers and other stakeholders in the financing, planning and conduct of research and technology transfer, with the view to enhancing the successful delivery of useful products and decision-support systems to farmers. A greater role is being played by the private sector in extension services.

The coming into force of several international treaties and conventions, notably the Convention on the Conservation of Biodiversity

©CAB INTERNATIONAL 1998. *Investment Strategies for Agriculture and Natural Resources* (ed. G.J. Persley)

and the Montreal Protocol to protect the ozone layer, and the establishment of the World Trade Organization has also increased the pressure on national governments to meet international obligations. Thus there have been several forces that have resulted in the restructuring, downsizing and refinancing of agricultural and natural resources research systems in both industrial and developing countries. These changes in domestic support for R&D are also having a significant impact on donor policies and financial support for national and international agricultural research.

The Findings

Setting policies

Possible principles for guiding government interventions in R&D are to:

- optimize the total public investments in R&D and their mix while minimizing the disincentives to private R&D;
- organize public sector R&D institutions so they can conduct R&D in the most cost-effective way, according to their comparative advantages with maximum synergies and minimum duplication;
- allocate and use resources efficiently among programmes and projects, minimize transaction costs, minimize administrative overheads and allow decentralized decision making, with effective incentives and accountability;
- minimize the transaction costs of raising revenue to finance public R&D by using least cost sources of funds.

Define and separate functions

Governments need to identify the roles and responsibilities for the separate functions of: setting policy for R&D funding; determining priorities; mobilizing finance; and managing R&D programmes and institutions, in order to achieve the desired outcomes from the development and delivery of knowledge, products and services to the stakeholders. Thus governments need to separate the decision-making functions on setting *policies* and *priorities* for R&D, from decisions on *public financing* of R&D programmes to *achieve desired outcomes*, from the *implementation* of research programmes by public and/or private institutions to *deliver agreed outcomes*. This separation provides clarity in responsibilities and accountability. Various agencies will specialize in their respective roles and responsibilities in setting policy, determining priorities, identifying needed outcomes, identifying most efficient means of achieving the outcomes and executing the research.

Determine priorities

A system must be established to determine priorities. Various options are available, including those based on economic models as well as more participatory systems. An effective system is likely to be a combination of needs analysis, participatory processes on priority setting involving a variety of stakeholders, input from economic models and other indicators, and informed judgements of stakeholders and decision-makers, leading to a consensus on priorities.

Focus on outcomes

There are desired outcomes for which public and/or private investors and potential beneficiaries are willing to pay for the necessary R&D in the belief that they will obtain a return on their investment. There needs to be a focus on the desired outcomes rather than on mobilizing a set percentage of agricultural GDP as a target amount to invest in R&D.

Identify new ways to mobilize finance

Governments need to establish the rationale and incentives for a mix of public and private sector funding of R&D, with taxation and other incentives for additional private sector investments; establish internal markets for R&D, based on customer/contractor relationships; use comparative advantage across the R&D system to minimize duplication; minimize transaction costs in mobilizing finance; and identify options for mobilizing finance and select those options most appropriate to achieve desired outcomes.

Effect and manage change

The following steps need to be taken when considering the reform process:

- Identify stakeholders: identify and involve stakeholders in the reform process.
- Examine options: undertake a *prior options* review to determine the scope and scale of reform required, at both system and institutional level.

- Identify change agents: identify and support the people who will be the change agents.
- Nurture the human resources: educate the people on whom the R&D system depends.
- Determine the time scale for change: set a realistic timetable for change to reduce uncertainties.
- Comparative advantage: identify comparative advantages, in order to maximize synergies and make the most efficient use of human and financial resources, within institutions, and at the national, regional and international level. Use subsidiarity as a principle.
- Financial incentives: provide financial incentives for change.

Create an enabling environment for R&D

The R&D system depends on inventive people and leaders with vision. This means providing an enabling environment for the conduct of science today and making science an attractive career for young people tomorrow. Universities and other tertiary institutions need to be actively involved in the R&D system both in the conduct of research and in the education of the next generation of researchers. An efficient and effective enabling environment for the conduct of R&D must be provided. This includes having in place effective regulatory systems which enable new products to reach the market; intellectual property regimes to enable access to new technologies and protection of inventions; information technology to access new knowledge globally; and fair taxation policies and other incentives for increasing private sector investments in R&D.

Access new knowledge/develop strategic alliances

New developments in biotechnology, information technology and global information systems are revolutionizing the conduct of R&D in agriculture and natural resources research. Science is now a global business. R&D systems need the best possible access to new knowledge in order to minimize duplication, apply the best available technologies to solve local problems and generate new knowledge which may have value locally, regionally and internationally.

Partnerships between national agencies and international groups (e.g. the research centres of the Consultative Group on International Agricultural Research) may often be the most effective means of developing and delivering new agricultural research technologies of a public goods nature. Natural resources management, which includes considerable 'public good' research appropriate for public investment, needs unique skills and new institutional arrangements, which may be best developed through strategic alliances. Environmental and natural resources management research issues are high on the international agenda. Natural resource management research presents challenges in moving from the required site-specific work to broader ecoregional application and impact. The use of satellite imagery, new models, global data systems and global information systems (GIS) technology will play an important role in linking these different levels.

The opportunities in biotechnology for agriculture and natural resources management are real and rapidly emerging. The developing world stands to benefit as well as the industrial world. Delivery of biotechnology products requires means and modalities for product development, intellectual property management, safe regulatory clearance procedures and production and marketing capabilities. Without these policies and practices in place, the technology and expertise will remain unused.

Identify ways of continually exchanging experience on reform of R&D systems

The options and experience of others should be examined and from this menu of choices those which are most applicable to the local situation should be selected. New developments in knowledge management should allow more ready access in the future to knowledge management systems which would present and update the experience of various countries in the reform of agricultural and natural resources research. This information would be a valuable tool for decision-makers.

Next Steps

Reform is a continuous process, and experience is being gained in many countries. It is important that those involved in reform have the opportunity to discuss and exchange with others their experiences and the lessons learned. The reform process needs to identify the key players, to determine what they need to know and identify the issues that need to be discussed under new investment strategies. These include: institutional flexibility, responsiveness to stakeholders, long-term effects of change and the need to get the reform process right. Change is not neccessarily always for the better. The process will also identify the potential interest groups and the menu of issues and options to be considered in embarking on reform.

The issues and options and key findings summarized here are elaborated further in the synthesis paper published by the World Bank

(Persley, 1998). The study produced a wealth of information on selected country and regional experience in the funding, organization and management strategies for agricultural research and natural resources management.

A series of background briefs targeted at different groups of stakeholders and decision-makers will summarize issues and options of most concern to each group. A synopsis of the material will be available on the Internet, through the World Bank website. The feasibility of developing a web-based knowledge management system to enable the experience in reform to be continually updated and experiences to be exchanged between various groups is being investigated further by the sponsors of the study in order to carry the study forward.

References

ABARE (1996) *Commodity Statistical Bulletin 1996*. Australian Bureau of Agricultural and Resource Economics, Canberra, Australia.

Albuquerque, R. and Salles-Füho, S. (1997) *Determinantes das Reformas Institucionais, Novos Modelos Organizacionais e as Responsabilidades do SNPA. Caracterizacao e Availacao das OEPAs–Relatorio Final*. Grupo de Estudios sobre Organizacaco da Pesquisa–GEOP, Departmento de Politica Cientifica e Technologica–DPCT/UNICAMP, Brazil.

Alex, G. (1996) *USAID and Agricultural Research: Review of USAID Support for Agricultural Research*. USAID Office of Agriculture and Food Security, Washington, DC.

Alston, J.M. and Mullen, J.D. (1992) Economic effects of research into traded goods: the case of Australian wool. *Journal of Agricultural Economics* 43, 268–278.

Alston, J.M. and Pardey, P.G. (1993) Market distortions and technological progress in agriculture. *Technological Forecasting and Social Change* 43 (3/4) (May/June), 301–319.

Alston, J.M. and Pardey, P.G. (1996) *Making Science Pay: the Economics of Agricultural R&D Policy*. American Enterprise Institute Press, Washington, DC.

Alston, J.M. and Pardey, P.G. (1998) The Economics of Agricultural R&D policy. In: Alston, J.M., Pardey, P.G. and Smith, V.H. (eds) *Paying for Agricultural Productivity*. IFPRI (forthcoming).

Alston, J.M., Anderson, J.R. and Pardey, P.G. (1995a) Perceived productivity, foregone future farm fruitfulness, and rural research resource rationalization. In: Peters, G.H. and Hedley, D.D. (eds) *Agricultural Competitiveness: Market Forces and Policy Choice, Proceedings of the Twenty Second International Conference of Agricultural Economists*. Aldershot, Dartmouth, pp. 639–651.

Alston, J.M., Norton, G.W. and Pardey, P.G. (1995b) *Science under Scarcity: Principles and Practice for Agricultural Research Evaluation and Priority Setting.* Cornell University Press, Ithaca, New York.

Alston, J.M., Sexton, R.J. and Zhang, M. (1997) The effects of imperfect competition on the size and distribution of research benefits. *American Journal of Agricultural Economics* 79(4), 1252–1265.

Alston, J.M., Marra, M.C., Pardey, P.G. and Wyatt, T.J. (1998a) *Ex Pede Herculem: a Meta-Analysis of Agricultural R&D Evaluations.* International Food Policy Research Institute, Washington, DC (forthcoming).

Alston, J.M., Pardey, P.G. and Smith, V.H. (eds) (1998b) *Paying for Agricultural Productivity.* International Food Policy Research Institute, Washington, DC (forthcoming).

Alves, E. (1992) *Getting Beyond the National Institute Model for Agricultural Research in Latin America, Case Study One, Agricultural Research in Brazil.* The World Bank, LAC Technical Department, Washington, DC.

Anderson, J. (1997) *On Measuring the Impact of Natural Resources Research.* Paper for UK Department for International Development Advisors' Conference, 9 July 1997, Sparsholt, UK (mimeo).

Anon. (1971) *A Framework for Government Research and Development* (Cm 4814). UK Government White Paper, London.

Anon. (1987) *Extension in Transition: Bridging the Gap Between Vision and Reality.* Report of the Futures Task Force to the Extension Committee on Organization and Policy, UK.

Anon. (1988) *Improving the Management of Government: the Next Steps.* UK Government White Paper, London.

Anon. (1991) *Higher Education: a New Framework* (Cm 1541). UK Government White Paper, London.

Anon. (1993a) *L'Experience Francaise du Developpement Agricole.* Association Nationale pour le Developpement Agricole, Paris, France, 30 pp.

Anon. (1993b) *Realising Our Potential – a Strategy for Science, Engineering and Technology* (Cm 2250). UK Government White Paper, London.

Anon. (1994) *Vision for the Future: a Strategic Plan for Agriculture.* USDA Co-operative State Research, Education, and Extension Service, Washington, DC.

Anon. (1995) *Managing for the Future: the CSREES Strategic Plan.* USDA Co-operative State Research, Education, and Extension Service, Washington, DC.

Anon. (1996) *Declaration and Plan of Action for Global Partnership in Agricultural Research.* Adopted at the Global Forum on Agricultural Research, October.

Antholt, C.H. (1994) *Getting Ready for the Twenty-first Century: Technical Change and Institutional Modernization in Agriculture.* World Bank Technical Paper No. 217. World Bank, Washington, DC.

Asian Development Bank (1983) *Review of the Bank's Role in Agriculture and Rural Development.* Board Paper R71-83, ADB, Manila, Philippines.

Asian Development Bank (1988) *Agricultural Research in the Asian and Pacific Region – Current Situation and Outlook.* Staff Working Paper, ADB, Manila, Philippines.

Asian Development Bank (1990) *A Review of Bank Support to Selected International Agricultural Research Centres.* Board Paper IN.270-90, ADB, Manila, Philippines.

Asian Development Bank (1992) *Annual Report.* ADB, Manila, Philippines.

Asian Development Bank (1995) *The Bank's Policy on Agriculture and Natural Resources Research.* Board paper R253-95, ADB, Manila, Philippines.

Atkinson, J.D. (1976) *DSIR's First Fifty Years.* Department of Scientific and Industrial Research, Wellington, New Zealand.

Bathrick, D.B., Byrnes, K.J., Stovall, J.G. and Podems, D.R. (1996) *Technology Institutions for Agricultural Free Trade in the Americas (TIAFTA).* LACTECH Project. USAID, Washington, DC.

Beyon, J. and Mbogok, S. (1996) *Financing of Agricultural Research and Extension for Smallholder Farmers in Sub-Saharan Africa.* Final Report. FSG, Oxford University Press, Oxford.

Birdsall, N. (1995) Proposal for the establishment of a regional fund for agricultural technology. In: *Proceedings of the Third Annual World Bank Conference on Environmentally Sustainable Development.* World Bank, Washington, DC.

Black, A.W. (1976) *Organizational Genesis and Development: a Study of Australian Agricultural Colleges.* University of Queensland Press, St Lucia, 316 pp.

Bosc, P.M. and Freud, E.H. (1994) Agricultural innovation in cotton zones of Francophone West and Central Africa: progress achieved and challenges ahead. Paper presented at the *IITA/FAO Workshop on Sustainable Cropping Systems for the Moist Savannah Zones, 19–23 September, 1994.* IITA, Ibadan, Nigeria.

Brennan, J.P. and Fox, P.N. (1995) *Impact of CIMMYT Wheats in Australia.* Economics Research Report No. 1/95. NSW Agriculture, Wagga Wagga, New South Wales.

Brenner, C. (1996) *Integrating Biotechnology in Agriculture: Incentives, Constraints and Country Experiences.* OECD Development Centre, Paris, 99 pp.

Brown, L.R. (1994) *Facing Food Insecurity.* State of the World Report, World Watch Institute. Washington, DC.

Burian, E. (1992) *Stated-owned Agricultural Research. A Comparative Study for France, the United Kingdom and Germany.* Agrarwirtschaft, Sh. No. 136. Pinneberg-Waldenau, Germany. (In German.)

Byerlee, D. and Alex, G.E. (1998) *Strengthening National Agricultural Research Systems: Policy Issues and Good Practice. Environmentally and Socially Sustainable Development, Rural Development.* World Bank, Washington, DC, 87 pp.

Byerlee, D. and Anderson, J.R. (1995) *Strategic Issues for Agricultural Research Policy in the 1990s.* World Bank,Washington, DC.

Byrnes, K. and Corning, S.L. (1995) *Programming for Sustainability: Lessons Learned in Organizing and Financing Private Sector Agricultural Research in Latin America and the Caribbean.* USAID, Bureau for Latin America, Washington, DC.

Cabinet Office (1994) *Efficiency Scrutiny of Public Sector Research Establishments* (Cm 2291), September 1995. HMSO, London.

Carney, D. (1998) *Changing Public and Private Roles in Agricultural Service Provision.* Overseas Development Institute, London.

Cashin, P.A. (1998) Is there any value in 'high value commodities'? *Australian Economic Papers* 7(2), 21–32.

CGIAR (1996a) *Report of the CGIAR Task Force on Central/Eastern Europe and the Former Soviet Union.* CGIAR Secretariat, World Bank, Washington, DC.

CGIAR (1996b) *Summary of Responses to the Survey of CGIAR Centers Conducted by the Private Sector Committee, December 1996.* CGIAR Secretariat, World Bank, Washington, DC.

CGIAR (1997a) *Financial Requirements of the 1998 CGIAR Research Agenda.* CGIAR Secretariat, World Bank, Washington, DC.

CGIAR (1997b) *The Role of the CGIAR in Biotechnology. Papers for the Midterm Meeting of the CGIAR, Cairo, May 1997.* CGIAR Secretariat, World Bank, Washington, DC.

CGIAR (1998a) *Report of a CGIAR Panel on General Issues in Biotechnology.* CGIAR Secretariat, World Bank, Washington, DC.

CGIAR (1998b) *Report of a CGIAR Panel on Proprietary Science.* CGIAR Secretariat, World Bank, Washington, DC.

Chambers, R.G. (1995) The incidence of agricultural policies. *Journal of Public Economics* 57 (2), 317–335.

Clarke, N.P. (1984) *Research 1984: the State Agricultural Experiment Stations.* Experiment Station Committee on Organization and Policy, Texas A&M University, College Station, Texas.

Clarke, N.P. (1993a) External and internal environments of the state agricultural experiment stations. Toward the 21st century: a multidimensional transition of the state agricultural experiment stations. In: *National Futuring Conference, June.*

Clarke, N.P. (1993b) Options for Improving Communications on Budget Proposals. Informal Working Paper, Texas A&M University, College Station, Texas.

Clarke, N.P. (ed.) (1996a) *From Issues to Action: a Plan for Action on Agriculture and Natural Resources for The Land Grant Universities.* National Association of State Universities and Land Grant Colleges, Board on Agriculture, Texas A&M University, College Station, Texas.

Clarke, N.P. (ed.) (1996b) *Opportunities to Meet Changing Needs: Research on Food, Agriculture, and Natural Resources. A Strategic Agenda for The State Agricultural Experiment Stations. January 1994* (updated April 1996). Texas A&M University, College Station, Texas.

Cohen, J.I. (1994) *Biotechnology Priorities, Planning, and Policies. A Framework for Decision Making.* ISNAR/IBS Research Report No. 6, ISNAR, The Hague.

Contini, E., Avila, A.F.D. and Reifschneider, F.J.B. (1997) Perspectivas de financiamento da pesquisa agropecuaria brasileira. *Cadernos de Ciencias e Tecnologia* 14(1), 57–90.

Corden, W.M. (1974) *Trade Policy and Economic Welfare.* Oxford University Press, Oxford.

Corning, S. (1993) *Organization and Financing of the Agricultural Research Programme of the Fundacion Chile: a Case Study.* USAID, LACTECH Project, Washington, DC.

Council for Research, Technology and Innovation (1997) *Biotechnology, Genetic Engineering and Economic Innovation – to Use Chances in a Responsible Manner.* Bonn, Germany. (In German.)

Crosson, P. and Anderson, J.R. (1993a) *Integration of Natural Resources and Environmental Issues in Research Agendas of NARS.* Briefing Paper No. 7, ISNAR, The Hague.

Crosson, P. and Anderson, J. (1993b) *Concerns for Sustainability: Integration of Natural Resource and Environmental Issues in Research Agendas of NARS.* ISNAR Research Report No. 4. International Service for National Agricultural Research, The Hague.

Dar, W.D. (1995) *Support for Agricultural Research System in South-East Asia: Impacts on Growth and Development.* Asia-Pacific Association of Agricultural Research Institutions (APAARI), FAO Regional Office, Bangkok, Thailand.

Doyle, J.J. and Persley, G.J. (1996*) Enabling the Safe Use of Biotechnology: Principles and Practice.* Environmentally Sustainable Development Studies and Monographs Series No. 10. World Bank, Washington, DC, 74 pp.

DTI (1996) *Biotechnology Means Business. Opportunities for Profits and Growth.* Information Pack, Department of Trade and Industry, London.

Dunn, E.G. (1997) *Basic Guide to Using Debt Conversion.* SD Publication Series. Technical Paper No. 44, USAID, Washington, DC.

Echeverría, R.G. (1990) Assessing the Impact of Agricultural Research. In: Echeverría, R.G. (ed.) *Methods for Diagnosing Research System Constraints and Assessing the Impact of Agricultural Research,* Vol.II, *Assessing the Impact of Agricultural Research.* ISNAR, The Hague.

Echeverría, R.G., Trigo, E.J. and Byerlee, D. (1996a) *Institutional Change and Effective Financing of Agricultural Research in Latin America.* Technical Paper No. 330. World Bank, Washington, DC.

Echeverría, R.G., Trigo, E.J. and Byerlee, D. (1996b) *Cambio Institucional y Alternativas de Financiacion de la Investigacion Agropecuaria en America Latina.* BID No. ENV-103, Inter-American Development Bank.

Ernst and Young (1995) *Biotech 96: Pursuing Sustainability. The Tenth Industry Annual Report.* Ernst and Young, Palo Alto, California.

Ernst and Young (1997) *Biotech 97: The Eleventh Industry Annual Report.* Ernst and Young, Palo Alto, California.

Ervin, D.E. and Schmitz, A. (1996) A new era of environmental management in agriculture. *American Journal of Agricultural Economics* 78 December, 1198–1206.

Ezaguirre, P. (1996) *Agriculture and Environmental Research in Small Countries. Innovative Approaches to Strategic Planning.* John Wiley and Sons, Chichester.

Falconi, C. and Elliott, H. (1995) Public and Private R and D in Latin America and the Caribbean. In: Peters, G.H. and Hedley, D.D. (eds) *Agricultural Competitiveness: Market Forces and Policy Choice.* International Association of Agricultural Economists, Dartmouth, UK.

Fan, S. and Pardey, P.G. (1992) *Agricultural Research in China – its Institutional Development and Impact.* International Service for National Agricultural Research, The Hague.

Fan, S. and Pardey, P.G. (1995) *Role of Inputs, Institutions, and Technical Innovations in Stimulating Growth in Chinese Agriculture.* EPTD Discussion Paper No. 13, IFPRI, Washington, DC.

FAO (1992) *Expansion of the CGIAR System*. TAC Secretariat, FAO, Rome.

FAO (1995) *Investment in Agriculture: Evolution and Prospects*. WFS 96/Tech/3, FAO, Rome.

Federal Ministry of Economic Cooperation (1992) (Federal Ministry of Food, Agriculture and Forestry): *Agricultural Research for the Tropics and Subtropics: Current Projects of Research Institutes in the Federal Republic of Germany*. Bonn, Germany.

Federal Ministry of Food, Agriculture and Forestry (1994) *Federal Research Centres and Blue-List-Institutes under the auspices of the Federal Ministry of Food, Agriculture and Forestry, 1994*. Bonn, Germany. (In German.)

Federal Ministry of Food, Agriculture and Forestry (1996) *Framework Concept for Federal Agricultural Research Adjustments*. Bonn, Germany.

Findlay, C.C. and Jones, R.L. (1982) The marginal cost of Australian income taxation. *The Economic Record* 58(162), 253–262.

Foley, K., Asimus, D., Gregson, T., Miller, J., Millis, N. and Shears, D. (1992) *CSIRO's Research for the Rural Industries: a Strategic Perspective*. CSIRO, Canberra, Australia.

Foster, A.D. and Rosenzweig, M.R. (1995) Learning by doing and learning from others: human capital and technical change in agriculture. *Journal of Political Economy* 103, 1176–1209.

Fuglie, K., Ballenger, N., Day, K., Klotz, C., Ollinger, M., Reilly, J., Vasavada, U. and Yee, J. (1997) *Agricultural Research and Development: Public and Private Investments Under Alternative Markets and Institutions*. Agricultural Economic Report No. 735. Economic Research Service, US Department of Agriculture, Washington, DC.

Goldsworthy, P., Ezaguirre, P. and Druiker, S. (1995) Collaboration between national, international, and advanced research institutes for eco-regional research. In: Bouma, J., Kuyvenhoven, A., Bouman, B.A.M., Luyten, J.C. and Zandstra, H.G. (eds) *Eco-Regional Approaches for Sustainable Land Use and Food Production*. Kluwer Academic, Dordrecht, The Netherlands, pp. 283–304.

Hamilton, A., Malcolm, J., Miller, R., Witters, R. and Bergmark, C. (1996) *Administrative Management Review: Sustainable and Natural Resources Management*. Collaborative Research Support Program, USAID, Washington, DC.

Harris, M. and Lloyd, A.G. (1991) The returns to agricultural research and the underinvestment hypothesis, a survey. *Australian Economic Review* 3rd Quarter, 16–27.

Hashim, M.Y. (1992) *The National Agricultural System in Malaysia*. Working Paper No. 41, International Service for National Agricultural Research, The Hague.

Hayami, Y. and Ruttan, V.W. (1985) *Agricultural Development: an International Perspective*. The Johns Hopkins University Press, Baltimore, Maryland.

HMSO (1995) *Government Response to the Efficiency Scrutiny of Public Sector Research Establishments* (Cm 2291), September 1995. HMSO, London.

Huang, S.-Y. and Sexton, R.J. (1996) Measuring returns to innovation in an imperfectly competitive market: application to mechanical harvesting of processing tomatoes in Taiwan. *American Journal of Agricultural Economics* 78, 558–571.

Huffman, W.E. and Evenson, R.E. (1993) *Science for Agriculture: a Long-term Perspective.* Iowa State University Press, Ames, Iowa.

Huffman, W.E. and Just, R.E. (1994) Funding, structure, and management of public agricultural research in the United States. *American Journal of Agricultural Economics* 76, 744–759.

Hullar, T.L. (1989) *Investing in Research: a Proposal to Strengthen the Agricultural, Food and Environmental System.* National Research Council, National Academy of Sciences, National Academy Press, Washington, DC.

Hussey, D.D. and Philpott, B.P. (1969) *Productivity and Income of New Zealand Agriculture 1921–1967.* Research Report No. 59, Agricultural Economics Research Unit, Lincoln College, Canterbury, New Zealand.

Indarte, E. (1997) Alianzas, flexibilidad y pensamiento estrategico en la organizacion y funcionamiento de la investigacion agropecuaria: el modelo INIA Uruguay. In: *Seminario Internacional Enfoques sobre Estrategias de Administracion y Financiamiento de la Investigacion Agropecuaria.* CIAT, Santa Cruz, Bolivia.

Industry Commission (1995) *Research and Development,* Vols 1–3. Australian Government Publishing Service, Canberra, Australia.

Inter-American Development Bank (1992) *El Rol de BID en el Fortalecimiento do la Investigacion Agropecuaria en America Latina y el Caribe.* Project Analysis Department, Inter-American Development Bank, Washington, DC.

Inter-American Development Bank (1996) *Towards a Regional System of Technology Innovation for the Food and Agricultural Sector.* Inter-American Development Bank, Washington, DC.

International Center for Agricultural Research in the Dry Areas (ICARDA) (1995) Assessment of research and seed production needs in dryland agriculture in the West and Central Asian republics: summary. In: *Proceedings of Workshop; 5–9 December, 1995; Tashkent, Uzbekistan.* ICARDA.

International Center for Agricultural Research in the Dry Areas (ICARDA). (1997) Highland Regional Programme: central Asia sub-region. In: *First Central Asia/ICARDA Regional Coordination Meeting. Aleppo, Syria. 13–16 September, 1997.* ICARDA.

ISNAR (1995) *Report of a Workshop: Research Policies and Management for Agricultural Growth and Sustainable Use of Natural Resources.* International Service for National Agricultural Research, The Hague.

Jaffe, W. and Infante, D. (1996) *Oportunidades y Desafios de la Biotecnologia para la Agricultura y Agroindustria de America Latina y el Caribe.* Department of Social Programs and Sustainable Development, Inter-American Development Bank, Washington, DC.

Jain, H.K. (1990) *Organization and Management of Agricultural Research in Sub-Saharan Africa: Recent Experience and Future Direction.* ISNAR Working Paper No. 33, The Hague.

Jain, H.K. (1995) *Agricultural Research Systems in South Asia: Organization and Management.* Asia-Pacific Association of Agricultural Research Institutions (APAARI), FAO Regional Office, Bangkok.

Jimenez, J. (1997) La experiencia Colombiana – la relacion sector publico–sector privado. In: *Seminario Internacional Enfoques sobre Estrategias de Administracion y Financiamiento de la Investigacion Agropecuaria.* CIAT, Santa Cruz, Bolivia.

Johnson, R.W.M. (1976) *Sector Accounts for Agriculture*. Research Paper No. 4/76, Ministry of Agriculture, Wellington, New Zealand.

Johnson, R.W.M. (1996) *Agricultural Productivity Trends for New Zealand 1972–1992*. Technical Paper No. 96/2. Ministry of Agriculture, Wellington, New Zealand.

Journeaux, P. and Stephens, P. (1997*) The Development of Advisory Services in New Zealand*. MAF Policy Technical Paper No. 97/8. MAF, Wellington, New Zealand.

Just, R.E. and Huffman, W.E. (1992) Economic principles and incentives: structure, management and funding of agricultural research in the United States. *American Journal of Agricultural Economics* 74, 1102–1108.

Kaimowitz, D. (1996) *La Investigacion sobre Manejo de Recursos Naturales para Fines Productivos en America Latina*. Inter-American Development Bank, Department of Social Programs and Sustainable Development, Washington, DC.

Kalaitzandonakes, N. (1997) *Commercialization of Research and Technology*. SD Publication Series, Technical Paper No. 43, USAID, Washington, DC.

Kenmore, P.E. (1996) *Integrated Pest Management in Rice*. In: Persley, G.J. (ed.) *Biotechnology and Integrated Pest Management*. CAB International, Wallingford, UK, pp. 76–97.

Lindarte, E. (1995) *Resultos del Inventario Institutional de 1993 sobre Recursos, Capacidad y Areas de Concentracion en Entidades de Investigacion Agropecuaria en America Latina y el Caribe*. IICA, San Jose, Costa Rica.

Lloyd, A.G., Harris, M.S. and Tribe, D.E. (1990) *Australian Agricultural Research: Some Policy Issues*. Crawford Fund for International Agricultural Research, Melbourne, Australia.

Lopez-Pereira, M.A. and Filippello, M.P. (1994) Maize Seed Industries Revisited: Emerging Roles of the Public and Private Sectors. In: *Maize 1993/94 World Maize Facts and Trends*. CIMMYT, Mexico.

Lovett, S.E. (1996) *A Corporate Conundrum: the Reforms to Rural Research and Development in Australia*. PhD Thesis, ANU, Canberra, Australia.

McIntire, J. (1995) Reviving the national agricultural research system of Mexico. In: Anderson, J.R. (ed.) *Agricultural Technology: Policy Issues for the International Community*. CAB International, Wallingford, UK.

McMahon, M. (1992) *Getting Beyond the National Institute Model for Agricultural Research in Latin America: a Cross-Country Study of Brazil, Chile, Colombia, and Mexico*. LAC Technical Department Regional Studies Program Report No. 20, World Bank, Washington, DC.

McMahon, M. (1996) *Rethinking Agricultural Research in Latin America*. World Bank, Washington, DC.

McMillan, W. (1997) *U.S. Interests in Economic Growth, Trade, and Stability in the Developing World. Conclusions and Recommendations of the Commission on International Trade, Development and Cooperation*. National Center for Food and Agriculture Policy, Washington, DC.

Miller, L. (1993) *Dynamics of the Research Investment: Issues and Trends in the Agricultural Research System*. USDA Cooperative State Research Service, Washington, DC.

MORST (1995a) *New Zealand Research and Experimental Development Statistics: all Sectors*. Ministry of Research, Science and Technology, Wellington, New Zealand.

MORST (1995b) *Priorities for 2001: Public Good Science Investment.* Ministry of Research, Science and Technology, Wellington, New Zealand.

MORST (1996) *New Zealand Research and Experimental Development Statistics: all Sectors.* Ministry of Research, Science and Technology, Wellington, New Zealand.

Moser, B. (1997) *Global Agricultural Science and Education Programs for America (Gasepa).* National Association of State Universities and Land Grant Colleges, Working Paper, April.

Mruthyunjaya and Ranjitha, P. (1997) Indian Agricultural Research Systems: Structure, Current Policy Issues, and Future Orientation. Draft Paper. Indian Council on Agricultural Research (ICAR), New Delhi.

Mueller, S. (1996) *How to Measure Sustainability: an Approach for Agriculture and Natural Resources.* Inter-American Institute for Cooperation on Agriculture, San Jose, Costa Rica.

Mullen, J.D., Lee, K. and Wrigley, S. (1996) Financing agricultural research in Australia: 1953–1994. In: Brennan, J.P. and Davis, J.S. (eds) *Economic Evaluation of Agricultural Research in Australia and New Zealand.* ACIAR Monograph No. 39, Canberra, Australia, pp. 45–57.

Mullenaux, J. *et al.* (1996) *Evaluation of the USAID Funded Collaborative Agricultural Research Networks in West and Central Africa.* USAID Draft Report, Washington, DC.

National Agricultural Technology Project (1995) *Rationalization of Agricultural Research in India; Executive Summary and Action Plan.* National Agricultural Technology Project, New Delhi.

New Zealand (1976, 1977, 1982, 1988, 1989) *New Zealand Official Yearbooks.* Department of Statistics, Wellington, New Zealand.

Nickel, J. (1997) *Institutional Issues Related to Agricultural Research in Pakistan.* World Bank, Washington, DC.

Nightingale, T. (1992) *White Collars and Gumboots: a History of the Ministry of Agriculture and Fisheries 1892–1992.* Dunmore Press, Palmerston North, New Zealand.

Noor, M.A. (1996) Successful diffusion of improved cash crop technologies. *Proceedings of the Workshop Developing African Agriculture: Achieving Greater Impact from Research Investments. Addis Ababa, Ethiopia, 26–30 September, 1995.* Sasakawa Africa Association, Mexico City.

NRC (1991) *Toward Sustainability: A Plan for Collaborative Research on Agriculture and Natural Research Management.* National Academy Press, Washington, DC.

NRC (1995) *Colleges of Agriculture at the Land Grant Universities: a Profile.* National Academy Press, Washington, DC.

Odegbaro, O.A. (1984) A case study: funding of agriculture in Nigeria. In: *Funding Agricultural Research in sub-Saharan Africa. Proceedings of a FAO/SPAAR/KARI Expert Consultation, Nairobi, Kenya, 1993.* FAO, Rome.

OECD (1997) *Environmental Indicators for Agriculture.* OECD, Paris.

Oehmke, J. (1997) *Summary of ex Post Rate-of-Return Studies for African Agricultural Technology.* Presented at Economic Impact Assessment Roundtable, USAID, Washington, DC.

Office of Sustainable Development (OSD)/USAID (1996) *Endowments in Africa.* SD Publication Series, Technical Paper, No. 24. USAID, Washington, DC.

OTA (1991) *Biotechnology in a Global Economy*, OTA-BA-494. US Congress, Office of Technology Assessment. US Government Printing Office, Washington, DC, 285 pp.

Pardey, P.G. and Alston, J.M. (1992) *Revamping Agricultural R & D*. 2020 Vision Brief 24, IFPRI, Washington, DC.

Pardey, P.G., Roseboom, J. and Anderson, J.R. (1991) *Agricultural Research Policy: International Quantitative Perspectives*. Cambridge University Press, Cambridge.

Pardey, P.G., Roseboom, J. and Beinteme, N.M. (1995) *Agricultural Research in Africa: Three Decades of Development*. ISNAR Briefing Paper No. 19. ISNAR, The Hague.

Pardey, P.G., Alston, J.M., Christian, J.E. and Fan, S. (1996) *Hidden Harvest: U.S. Benefits from International Research Aid*, IFPRI Food Policy Report. International Food Policy Research Institute, Washington, DC.

Pardey, P.G., Alston, J.M., Christian, J.E. and Fan, S. (1997a) A Productive Partnership: the Benefits from US Participation in the CGIAR. International Food Policy Research Institute, Washington, DC (draft report).

Pardey, P.G., Roseboom, J. and Beinteme, N.M. (1997b) Investments in African agricultural research. *World Development* 25(3).

Pardey, P.G., Roseboom, J. and Craig, B.J. (1998a) Agricultural R&D Investments and Impact. Chapter 3. In: Alston, J.M., Pardey, P.G. and Smith, V.H. (eds) *Paying for Agricultural Productivity*. IFPRI (forthcoming).

Pardey, P.G., Roseboom, J. and Fan, S. (1998b) Trends in Financing Asian and Australian Research. In: Tabor, S.R., Janssen, W. and Bruneau, H. (eds) *Financing Agricultural Research: a Sourcebook*. International Service for National Agricultural Research, The Hague, pp. 341–356.

Parr, J., Papendick, R., Youngberg, I. and Meyer, R. (1990) Sustainable agriculture in the United States. In: Edwards, C.A., Lal, R., Madden, P., Clive, R.H. and House, G. (eds) *Sustainable Agricultural Systems*. Soil and Water Conservation Society, Ankeny, Iowa.

Pearce, A. (1995) Contracting in the science sector: a research provider's view. *Public Sector* 18(4), 10–15.

Penning de Vries, F., van Keulen, H. and Rabbinge, R. (1995) Natural resources and limits of food production in 2040. In: Bouma, J., Kuyvenhoven, A., Bouman, B.A.M., Luyten, J.C. and Zandstra, H.G. (eds) *Eco-regional Approaches for Sustainable Land Use and Food Production*. Kluwer Academic, Dordrecht, pp. 283–304.

Persley, G.J. (1990a) *Beyond Mendel's Garden: Biotechnology in the Service of World Agriculture*. CAB International, Wallingford, UK.

Persley, G.J. (1990b) *Agricultural Biotechnology: Opportunities for International Development*. CAB International, Wallingford, UK.

Persley, G.J., Giddings, L.V. and Juma, C. (1992) *Biosafety: the Safe Application of Biotechnology in Agriculture and the Environment*. World Bank/International Service for National Agricultural Research, The Hague, 39 pp.

Persley, G.J. (1998) *Investing in Knowledge for Development: New Strategies for Agriculture and Natural Resources Research Systems*. Environmentally and Socially Sustainable Development Studies and Monographs Series, World Bank, Washington, DC.

Petit, M. (1996) *The Emergence of a Global Agricultural Research System: the Role of Agricultural Research and Extension Group (ESDAR)*. World Bank, Washington, DC.

PHARMA (1996) *Biotechnology in Development Medicines 1996 Survey*. Pharmaceutical Research and Manufacturers of America, Washington, DC, 28 pp.

Posada, R. (1992) *Getting Beyond the National Institute Model for Agricultural Research in Latin America, Case Study Three, the Agricultural Research System of Colombia: a Review*. LAC Technical Department, World Bank, Washington, DC.

Powell, M.J. (1995) Lessons Learned from Networking: Organization, Management and Costs. Agricultural Research Networks of Eastern, Central and Southern Africa. USAID/SPAAR, Washington, DC (unpublished).

Pray, C. (1997) *Final Report on the China Agricultural Research Funding Project*. Report to the World Bank, Washington, DC.

Preuss, H.A. (1996) Agricultural Research and Sustainable Management of Natural Resources – Introductory Notes and Summary of Contributions. In: H.J. Preuss (ed.) *Agricultural Research and Sustainable Management of Natural Resources*. Centre for Regional Development Research, Justen-Liebig-University, Giessen, Germany.

Pritchard, A.J. (1990) *Lending by the World Bank for Agricultural Research, a Review of the Years 1981 through 1987*. Technical Paper No. 118. World Bank, Washington, DC.

Runge, C. (1992) A policy perspective on the sustainability of production environments: toward a land theory of value. In: *Future Challenges for National Agricultural Research: a Policy Dialog: Proceedings of an International Conference*. ISNAR, The Hague.

Sarles, M. (1990) USAID's experiment with the private sector in agricultural research in Latin America and the Caribbean. In: Echeverría, R.G. (ed.) *Methods for Diagnosing Research System Constraints and Assessing the Impact of Agricultural Research*, Vol. 1. ISNAR, The Hague.

Schedvin, C.B. (1987) *Shaping Science and Industry: a History of Australia's Council for Scientific and Industrial Research, 1926–1949*. Allen & Unwin, Sydney, 374 pp.

Schiff, M. and Valdez, A. (1992) *The Plundering of Agriculture in Developing Countries*. World Bank, Washington, DC.

Scientific Council (Wissenschaftsrat) (1992a) *Recommendations on the Future Structure of Universities in East Germany and in Eastern Berlin, Part II* (in German). Köln, Germany, 328 pp.

Scientific Council (Wissenschaftsrat) (1992b) *Evaluation Report Concerning Research Institutions Outside Universities of the Former GDR in the Field of Agricultural Sciences*. Köln, Germany, 455 pp. (In German.)

Scobie, G. and Eveleens, W. (1986) *Agricultural Research: What's it Worth?* Discussion Paper 1/86. Ministry of Agriculture and Fisheries, Hamilton, New Zealand.

Scott, R.B. (1992) Scientific advances and agricultural technologies as opportunities for NARS: the case of forestry and agroforestry. In: *Future Challenges for National Agricultural Research: a Policy Dialog: Proceedings of an International Conference*. ISNAR, The Hague.

SPAAR Newsletter (1996) The Contract Research Programme for Agriculture in Malawi. In: *Contact*. SPAAR, World Bank, Washington, DC.

SPAAR Secretariat (1996) *Special Program for African Agricultural Research (SPAAR) Annual Report 1996*. World Bank, Washington, DC.

Springer-Heinze, A. (1996) The nexus between natural resources management and public policy making. In: Preuss, H.J. (ed.) *Agricultural Research and Sustainable Management of Natural Resources*. Centre for Regional Development Research, Justen-Liebig-University, Giessen, Germany.

SPRU (1996) *The relationship between Publicly Funded Basic Research and Economic Performance*. (Science Policy Research Unit, University of Sussex.) HM Treasury, London.

Spurling, A. *et al.* (1992) *Agricultural Research in Southern Africa: a Framework for Action*. World Bank Discussion Paper, Africa Technical Department Series No. 184. World Bank, Washington, DC.

State Government of Victoria (1994) Submission to the Industry Commission Inquiry into Research and Development in Australia. Department of Agriculture, Melbourne, Australia (mimeo).

STEP Report (1992*) Science and Technology Expert Panel. Investing in Science for our Future*. Ministry of Research, Science and Technology, Wellington, New Zealand.

Stewart–Levene Report (1992) *Review of Allocation, Management and Use of Government Expenditure on Science and Technology*. UK Government White Paper, London.

Summers, R. and Heston, A. (1991) The Penn World Trade (Mark 5): An Expanded Set of International Comparisons, 1950–1988. *The Quarterly Journal of Economics*, May.

Tabor, S.R. (ed.) (1995) *Agricultural Research in an Era of Adjustment. Policies, Institutions, and Progress*. Economic Development Institute Seminar Series. World Bank, Washington, DC.

Tabor, S. and Alirahman (1995) Indonesia. In: Tabor, S. (ed.) *Agricultural Research in an Era of Adjustment: Policies, Institutions, and Progress*. EDI Seminar Series. World Bank, Washington, DC, pp. 85–100.

Tabor, S.R. and Ballantyne, P. (1995) *Structural adjustment and agricultural research*. Economic Development Institute, World Bank, Washington, DC.

Tabor, S.R., Janssen, W. and Bruneau, H. (1998) Financing Agricultural Research: A Sourcebook. ISNAR, The Hague.

TAC (1996) *A Strategic Review of Natural Resources Management Research on Soil and Water*. TAC Secretariat/CGIAR, FAO, Rome.

TAC (1997) *Financial Requirements of the 1998 CGIAR Research Agenda*. TAC Secretariat/CGIAR, FAO, Rome.

Taylor, Ajibola *et al.* (1996) *Strengthening National Agricultural Research Systems in the Humid and Sub-humid Zones of West and Central Africa; a Framework for Action*. World Bank Technical Paper No. 318. World Bank, Washington, DC.

Tribe, D.E. (1991) *Doing Well by Doing Good*. Pluto Press Australia Ltd, Leichardt, in association with The Crawford Foundation for International Agricultural Research, Parkville, Australia.

Trigo, E.J. (1995) *Agriculture, Technological Change, and the Environment in Latin America: a 2020 Perspective. Food Agriculture and the Environment*.

Discussion Paper 9. International Food Policy Research Institute, Washington, DC.

Umali, D.L. (1995) *Public and Private Sector Roles in Agricultural Research. Theory and Experience*. World Bank Discussion Paper No.176. World Bank, Washington, DC.

Upton, S. (1995) Contracting in the science sector: an overview. *Public Sector* 18(4), 2–5.

Venezian, E. (1992) *Getting Beyond the National Institute Model for Agricultural Research in Latin America, Case Study Two, Agricultural Research in a Growing Economy: The Case of Chile 1970–90*. LAC Technical Department, World Bank, Washington, DC.

Venezian, E. and Muchnik, E. (1995) Structural adjustment and agricultural research in Chile. In: Tabor, S.R. (ed.) *Agricultural Research in an Era of Adjustment Policies, Institutions, and Progress*. Economic Development Institute Seminar Series. World Bank, Washington, DC.

Venner, R.J. (1997) An Economic Analysis of the U.S. Plant Variety Protection Act: the Case of Wheat. PhD dissertation, University of California, Davis, California.

Waage, J. (1996) Integrated pest management and biotechnology: an analysis of their potential for integration. In: Persley, G.J. (ed.) *Biotechnology and Integrated Pest Management*. Biotechnology in Agriculture No. 15, CAB International, Wallingford, UK, pp. 37–60.

Walsh, J. (1991) *Preserving the Options: Food Productivity and Sustainability – Issues in Agriculture No. 2*. CGIAR, World Bank, Washington, DC.

Weijenberg, Jan *et al.* (1993) *Revitalizing Agricultural Research in the Sahel: a Proposed Framework of Action*. World Bank Discussion Paper, Africa Technical Department Series No. 211. World Bank, Washington, DC.

Weijenberg, Jan *et al.* (1995) *Strengthening National Agricultural Research Systems in Eastern and Central Africa: a Framework for Action*. World Bank Technical Paper No. 290. World Bank, Washington, DC.

Winkler, D.R. (1990) *Higher Education in Latin America: Issues of Efficiency and Equity*. World Bank Discussion Paper No. 77. World Bank, Washington, DC.

World Bank (1992) *Public and Private Sector Roles in Agricultural Research*. Discussion Paper No. 176. World Bank, Washington, DC.

World Bank (1995a) *World Tables, Diskette version*. World Bank, Washington, DC.

World Bank (1995b) *Agriculture and Poverty. Agricultural Symposium, 5–6 January 1995* (various documents). World Bank, Washington, DC.

World Bank (1996) *Strategic Issues for Agricultural Research Policy to 2000 and Beyond*. Agricultural and Natural Resources Department and ESDAR, World Bank, Washington, DC.

World Bank (1997a) *Kazakstan: Transition of the State: a World Bank Country Study*. World Bank, Washington, DC.

World Bank (1997b) *Project Appraisal Document on a Proposed Loan of US$60 Million Equivalent to the Federative Republic of Brazil for an Agricultural Technology Development Project*. Report No. 16520-BR, 5 May, 1997. World Bank, Washington, DC.

Wright, B.D. (1983) The incentives to invent: patents, prizes and research contracts. *American Economic Review* 73 (September), 691–707.

Wyke, A. (1988) The genetic alternative: a summary of biotechnology. *The Economist* 30 April 1988.

Index

Figures in **bold** indicate major references.
Figures in *italic* refer to diagrams, photographs and tables.